for John Linder —
may he be a
friend forever.

MODELS OF MAN

MODELS OF MAN

A PHENOMENOLOGICAL CRITIQUE OF SOME PARADIGMS IN THE HUMAN SCIENCES

by

JAMES J. DAGENAIS

MARTINUS NIJHOFF/THE HAGUE/1972

ISBN 90 247 1290 4

PRINTED IN BELGIUM

TABLE OF CONTENTS

For My Parents

PREFACE

This essay is, first, a theoretical and historical study of some classical scientific ways of studying human being in the world. The more readily accessible and more commonly discussed "models" of being human were chosen for review here, but structuralism is included because I believe it will have the same impact in America as it has had in France, and I hope that American readers might be forewarned about what may be ideologically at stake before the technical, and fruitful, aspects of the movement become an academic fad in the United States. The subjects included are mainline experimental psychology from Wundt to Skinner, with its relatively shortlived functionalist and Watsonian-behaviorist formulations; holistic psychology from Brentano through Stumpf, Husserl, and Goldstein to Maslow, Rogers, and contemporary "third force" psychology; and the psychoanalytic model, for which the only paradigm is Freud himself. Preeminence is given to psychological paradigms, since their subject matter lies closest to the classical philosophical tradition from which "philosophical anthropology" emerged. (This book is, in the final analysis, a prolegomenon to an articulated philosophical anthropology.) Sociological models are also considered: the "classical" tradition from Comte to the present, and Marxist anthropology from the manuscripts of 1844 to the present. The structuralist model, from Durkheim to Chomsky, is also considered, since it cuts across and gives new dimensions to all the foregoing models.

The essay is, second, a phenomenological critique of these historico-theoretical considerations. The thesis maintained is that the human sciences, as sciences, must attempt to reduce the meaning of man to the control of the scientific presuppositions which found each science, and that, in consequence, each scientific model can and must pretend to universal exclusiveness. Furthermore, since each science must be limited to one perspective, they cannot all be summed up under the control of another science, such as philosophy. This amounts to saying that the sciences (positive, axiomatic, or humanistic) must be autonomous as sciences; that the only critique of them *as sciences* is from within the sciences themselves. Any other knowledge we have of human beings outside of these sciences is, in respect to them, unscientific. Put in phenomenological terms, Husserl maintained that "every originary given intuition is a legitimate source of cognition, that everything which presents itself to us in originary intuition (in its bodily actuality, so to speak) is to be taken just as what it gives itself, but only within the limits in which it gives itself." The latter part of this principle of all principles indicates that, phenomenologically speaking, each science is a noetic activity directed towards one and the same noema—human being. The same thing could be said in classical Scholastic terms: that human being is the "material object" of the human sciences, and that each of them attains its object under different formal aspects.

The thesis, then, first defends the human sciences as legitimate sources of cognition; but then must come to terms, in a second phase, with the lack of agreement about human beings in the world among the sciences which study them. The problem here lies, really, on the *noetic* side of the enterprise—not in the formulations of the sciences, but in the formulations by scientists. The *presuppositions* of the sciences of man are radically different and often contradictory; and after all we philosophers are as interested in knowing as much as we are in the known. If this is so, how can we answer the question about how we *know* human being?

First, as noted above, we might hypothesize that we know man through the sciences of man. But, again, each of these sciences is automomous and independent, given its presuppositions and operational or other methods. Reflection upon the sum of the results of the human sciences will not reveal what human being is except that, to use Aristotle's dictum, "it is said in many ways." Nor will such reflection reveal how we know what human being is except that we know him in many scientifically irreconcilable ways.

Second, we might hypothesize that we know man through a "definition" of man. But then the elaboration of an all-encompassing theory about human being would have to depend upon all the empirical sciences anyway. Otherwise it would have only the apodicticity of a logically necessary statement. That is, if it is to be about real human beings, such a theory will have to depend upon a host of extra-systematic assumptions which will serve only to invalidate the supposed logical consistency of the argument.

Third, we might hypothesize that we know through a prestructured "metaphysical system" of the whole. This has the ring of dishonesty, in that the empirical evidence will be sifted to provide "examples" of what is already known in a presystematized way. This really makes the sciences of man unnecessary and gratuitous, and explains nothing about the origin of the system in any case.

In all these inadequate hypotheses there is one recognizable constant: that all understanding of human being in the world, whether scientific or philosophical, is founded upon a pre-scientific and pre-philosophical experiencing of human being as self-and-other-in-the-world. The only alternative, then, is a critical explicitation of this experiencing; and that is the task of philosophy. It is the task undertaken in this essay, especially in the important and basic defense of the second phase of the thesis stated above.

A corollary to this thesis is that there are divergent trends not only among the human sciences but within each of them—trends due to the ambiguities built into their scientific foundations by their founders and paradigm-workers. Ultimately, then, all scientific, philosophical, or everyday explicitation of our experiencing in the world is based upon fundamentally free options.

A second corollary is thus imposed which is of perennial interest but of secondary philosophical importance: Whatever conclusions scientists reach regarding human freedom in relation to scientific determinism cannot be validly reached so long as their "originary given intuition" is taken as "just what it gives itself to be" and "only within the limits in which it gives itself;" for the limits within which the given is given to a science are predetermined by the kind of "fact" which scientists design their science to deal with. That all science is and must be "objective" is the first and most serious barrier to the understanding of human beings; for "objectivity" is already a second look. The objectivizing consciousness is characterized by a decision to render the world, and human being in the world, an object. As Ronald Laing has pointed out, the facts with which science deals are not "data" but "capta." The explicitation of human "freedom," then, in terms of scientific "causality" is another instance of the extension of a fundamentally free, non-scientific option.

Before getting on with the lonely one-man interdisciplinary task which lies ahead, I should like to express sincere gratitude to Charles Hampden-

Turner, author of *Radical Man.* Charles generously read the working draft of this essay, and demonstrated his empathic genius to know sometimes better that I what I am up to in these pages. When one attempts to explicitate the foundations of one's own thought and one's own options, one is bound to be halting, unclear, and often turgid. It is regrettable that Charles cannot force his own lucidity upon another. The reader is urged to look for Hampden-Turner's forthcoming article, "The Hidden Moralities of Social Science;" it is a masterful application of the thesis which is here so inchoatively upheld.

CHAPTER I

THE PSYCHOLOGICAL MODEL

THE "SCIENTIFIC" REVOLUTION AND REAR-GUARD PHILOSOPHICAL ACTION

We shall eventually come to an evaluation of "the present status of philosophical anthropology;" but the reader may find it more than a little odd that the enterprise had its beginning in Germany 1874. That was the year, however, in which both Wundt's *Grundzüge der Physiologischen Psychologie* and Brentano's *Psychologie vom empirischen Standpunkt* were published. Before them, of course, psychological reflection had abounded, in various forms: Helmholz and Fechner were immediate predecessors of Wundt; and Bain, Spencer and Mill (and also Kant, Herbart and Lotze) figure in the background, while Brentano's inspiration was drawn from the observational penchant of Goethe and Hering (and, more basically, from Aristotle). But Wundt is considered by all historians as the first modern experimental psychologist, and Brentano as the father of modern phenomenology. Boring draws the contrasts between these two enterprises: "Brentano's psychology was empirical but not experimental; Wundt's was experimental. Hence Brentano's method was argumentative and Wundt's was descriptive in intention . . . Brentano organized his system about the psychical act; Wundt built his about sensory contents." Between the two there is a world of difference, although both were attempting to formulate the "new" psychology of the late nineteenth century as a *science;* and Boring underlines the fifty-year dilemma of systematic psychology as the choice between Brentano and Wundt—especially between act psychology and content psychology. Only Külpe managed to reconcile the two currents, and then only briefly. The famous debate between McDougall and Watson in the 1920's replayed the original scenario created fifty years earlier. The Rice symposium on *Behaviorism and Phenomenology,* in 1964, featuring a confrontation between Carl Rogers and B.F. Skinner was simply a re-enactment of the century-old script, (the dress rehearsal had been held in 1956)[1] and the debate continues.

Only a few years ago, one might have concluded that behaviorism was on the way out; the prestigious Sigmund Koch was ready to laugh it out of existence, and proposed that "the position be considered once and for all defunct." And yet, in our universities, in our marketing studios, in our urban planning centers, in almost every sector of our public and private lives, the humanist still bows to the experimenter, whose demand for "more hard facts," "more computer time," "more rationalization" receives priority attention, research funding, prestige, and obedience. The consequences of these conflicting points of view will be drawn more concretely in succeeding chapters. For the moment, let us dwell upon the project of the founding fathers of modern psychology, and understand why Skinner's operationalism (which the author considers profoundly anti-human) came to be the contemporary spokesman of the original enterprise.

Wilhelm Wundt (1832-1920) received his early training in medicine, and was therefore much inclined to respect the importance and fruitfulness of experimentation; even more important for the present context is the fact that many years of experimentation in sense physiology had preceded him. (One thinks of Weber, Fechner, Helmholz and Herbart.) In 1856, his doctorate in medicine in hand, he became professor of physiology in Heidelberg, with, somehow, a conviction born of critical reflection upon the work of his predecessors, that "perception" is more than what the physiologists could describe as "sensation." His views were large, and his lectures centered not simply upon descriptive anatomy and physiology, but also upon epistemology and upon anthropology as the natural history of man.

His first really important book, the *Beiträge zur theorie der Sinneswahrnemung,* appeared in three parts in 1858, 1861, and 1862; and Tichener's later assessment of it was that it programmed Wundt's entire prolific and extensive life's work. Basic to the program of "experimental psychology" which it mapped out was the idea of *experimentation.* This was only natural, since, if psychology is to be a science in the true sense of the word, it must of course use a scientific method; and experimentation is the method of science. This was the opinion of Fechner; and Wundt followed him in it instead of following the opinion of Herbart

[1] See *Science*, 1956 (124), No. 3231, 1057-1066, for a report of Skinner's and Rogers' Symposium on "Some Issues Concerning the Control of Human Behavior."

(for whom experimentation was not necessarily *the* scientific method in psychology as it was for physics), although it was to Herbart that the idea of a scientific psychology was credited.

Though Wundt accepted a chair of philosophy in Leipzig in 1875, he was also given the world's first laboratory in experimental psychology. Whether Wundt was himself an experimentalist or not need not be discussed here; it is enough to note that his students were, and that they, under his inspiration, or at least under his aegis, created the impressive collection of psychological experiments which characterized the Leipzig enterprise and influenced, in one way or another, the scientific psychology of the late 19th and early 20th centuries, and still continues to haunt experimental psychology today.

From 1880 on, Wundt's encyclopedic philosophical interests came to the fore, and his books on Logic and Method, on Ethics, and a system of scientific philosophy, or metaphysics, were his main preoccupations. Whether he participated or not in the elaboration of scientific projects in experimental psychology is not the important factor here; the philosophical background of experimental psychology was his work. Though his ideas were not as "mechanical" and "elementist" as his critics (for example McLeod) have tended to think, yet the flaws in his system were enough to provoke the later behaviorist "correction" as well as the functionalist revolt and the continued hostility of the "act" psychologists, including the early (and contemporary) phenomenologists. Wundt was responsible for the idea that psychology as a science is not a science of "inner experience;" for, as Boring explains, Wundt was convinced that

> the distinction between inner and outer experience is not valid; feeling is 'inner' in that it is consciously subjective, and perception is 'outer' in that it refers to objects, but psychology deals with both alike . . . Psychology deals, not with inner experience, but with *immediate experience* . . . [2]

Thus, if psychology's subject is immediate experiencing, its method must be the same, since, as Boring further explains, "the subject-matter and the method of psychology cannot be discussed separately." [3]

This means that Wundt's psychological experimenting was basically a science of introspection, in an effort to describe "the mind" in terms of its formal elements, which are psychologically compounded through association. Introspection, in the Wundtian enterprise, was an effort at analysis of mind into its elements and an effort to understand how these

[2] Edwin G. Boring, *A History of Experimental Psychology*, 2nd ed. New York, 1957, 332.
[3] *Ibid.*

elements are compounded in experience. The science of psychology continued to be the science of introspection in these terms until "objective" psychology spotted the flaw in Wundt's option. For Wundt was a psychological dualist; in spite of his articles on the physiology of the central nervous system, and so on, his tendency was to leave "the body" more and more out of consideration, and to espouse a kind of psychophysical dualism inherited from a tradition even earlier than Descartes.

What provided the motor force of scientific psychology under the tutelage of Wundt was his method of analysis—the analysis of "mind" (of conscious processes) into simple quantities, and the analysis of the laws of their interconnectedness. Thus, "elementism", and the consequent atomization of "mind," and later of "behavior," was the legacy of Wundt.

It is impossible here to detail the enormous body of data and the famous pioneers of experimental psychology who contributed to its formation. For these details the reader must consult the histories of experimental psychology and the writings of the experimenters themselves. The effort here is to elucidate the philosophical background of the project of scientific psychology in its origins and its development. The over-simplification of the history of experimental psychology, and its critique, is regrettable; but like the archaeologist, the philosopher is constrained to leave something for future more detailed examination.

The following pages can be outlined rather simply: Wundt's psychology was marked by an "elementism" and a penchant for the study of "consciousness." The first revolt within the school of Wundt refused the "elementism" but kept "consciousness;" the second kept "elementism" and rejected "consciousness." We shall briefly trace these successive movements within the tradition of Wundt, and contrast them later with the tradition of Brentano.

The first revolt against the "general psychology," in the tradition of Wundt, was the American psychology of function. Boring notes, in discussing the functional psychology of America, that functionalism has a particularly American cast, insofar as the American temper fostered it, and insofar as functionalism is close to common sense. He notes, more generally, and with great acumen, that:

> The alternative to function is structure. If, influenced by culture, you conclude that you should devote yourself to the description of nature, content to say *what* happens and *how* it happens, without asking the question why, then you are concerned with structure, are working in the descriptive tradition of the taxonomists and, in psychology, the tradition of Wundt, Külpe and Tichener. But if you ask *why,*

if you try to understand causes, then you are interesting yourself in capabilities, in functions, and are being a functionalist. [4]

It follows from this that functionalism tends toward prediction, toward the future, toward project, change and control. In America, with its psychological roots both in Germany and in England, experimentation was to be concerned less with what had been (a static artifact) that with what might be. For American psychology, in addition to Wundt, there was Darwin, whose theory of evolution interested many, of course, in what man had become, in view of his origins; for many more, it represented the challenge of what man might yet become. Functional psychology takes many forms, and it is again regrettable that the modalities of these forms cannot be explicitated here. We shall have to be content with Boring's broad and masterful characterization, leaving out also the many steps between Darwin and Dewey, as well as those between Wundt and Dewey. In the final analysis, John Dewey serves here as no more than a figurehead, a standard-bearer of what was, after all, a very ancient way of understanding man, as well as a by no means homogeneous movement in psychology, even in America. Dewey was the focal point and organizing principle of functional psychology at Chicago, and later, with Cattell, at Columbia, where Edna Heidbreder experienced the freedom from systematic restraints which characterized the whole movement at its best.

In spite of its many ramifications and developments, the functionalism which opposed Wundt and Tichener might be called molar, as against the earlier elementism. Dewey's paper, "The Reflex Arc Concept in Psychology" published in 1896, took a clear stand against the reigning concept of elementism, or atomism, not so much in the area of sensation as in that of the reflex concept which is at the heart of the behavioral counterpart of atomistic sensation. In sum, setting his attack against the simple S-R concept of "reaction," and even the more sophisticated idea of concatenations of S's and R's, Dewey insisted that we are dealing with total reactions, total coordinations of "responses" which are "strictly correlative and contemporaneous" with stimuli, to the extent that nothing can be called a "reaction" except in function of a "stimulus," and nothing can be called a stimulus except insofar as there is a response to it. Furthermore, he insisted, the coordination of a total response is adaptive and purposeful for the person acting. Thus, two strands of the original functionalist revolution, the molar concept of coordinations (later to be

[4] *Ibid.*, 551.

recognized by Goldstein and incorporated into an even older evolutionary biological and sociological theory) were present in Dewey's paper of 1896. However, since functional psychology lends itself, as a theory, to *applied* psychology, in view of the fact that, functionally, there is *something to do,* the offshoot of Dewey's insight was a centering of interest in new applications: animal psychology got a new lease on life, and so did abnormal psychology and its concomitant therapy. Social psychology is another offshoot of functionalism, and so are all the twentieth century psychologies: child psychology, vocational counseling, clinical psychology, educational, and industrial psychology. It must be noted, however, that the "functionalist," in the broad sense noted above, shares *all* these fields with the "taxonomists," and presently takes a back seat to them. Dewey himself, as a philosopher rather than an experimentalist, was a philosopher of social change and progress, and became noted more for his democratic evolutionism, as translated into the philosophy of education, itself integrated into the American philosophy of pragmatism.

Would it have been better, from a humanistic point of view, if Dewey had been more interested in the non-reaction of the scientist to his paper on the Reflex Arc Concept, and had continued developing those ideas? Perhaps he would never have become the Grand Old Man of American Educational Philosophy if he had done so; but then the field was left to Cattell and Thorndike, to Watson, Holt, Tolman and Lashley. One may see in these remarks a tendency to downgrade the work of the great pioneers in experimental psychology; but such is not the intention. However, it would have been wiser, it seems to me, if American psychology had had the benefit of corrective *philosophical* thought, over and above the internal theoretical corrective of students, masters, and colleagues. American psychology grew up talking to itself, in mechanistic and atomistic terms; when it was finally asked to speak to the populace, it found no distinction to speak of between the human being and the laboratory rat; and requests for marketing studies, and other "how to" recipes are filled in terms of operant conditioning. This is to say that the original functionalist manifesto has somehow been directed to the Dead Letter Office; the postal functionary has delivered instead notice of the operational and structuralist counter-revolution.

Of course, there was a question to be asked and answered, and a choice to be made, rather early in the game. The question is, "Does psychology deal with consciousness, or with behavior or both?" As early as Locke and Berkeley, and as late as Wundt, psychology was the study of *consciousness*; experimental psychology was thought to be, for a long time, the study of conscious acts. In an empirical, but not experimental context, that of Brentano and the phenomenologists, psychology was also an investigation of consciousness. Within these groups, there were subgroups which tended toward a physiological psychology, to be sure, and attempted to trace the neurological events underlying conscious phenomena—as much in the tradition of Wundt as in the isomorphism of the Gestaltists.

The opposite interest has been shown, if not as early, by students of behavior, let us say, by Pavlov and Watson. The present fashion is to translate the givens of an hypothetical consciousness into the givens of behavior (of speech or action) as observable. The interiority of the subject is understood to be inaccessible, and his report of it unreliable. Better to substitute for the supposed "object of observation" (who is probably lying anyway) his observable operations, by means of the operational reduction.

How did this come about? From the ancient science of animal psychology, of course. The observation of animal behavior and theories regarding it have a long history which cannot be retraced here; but it is evident that, even if Dr. Lorenz can talk to the animals, it is not sure that they can talk to him. In modern times, at least in America, it is customary to date experimental animal psychology from 1898, the date of Thorndike's report on animal intelligence. "Comparative psychology" was a logical next step. If one wanted to study the mind in evolution, why not begin with a more primitive (animal) mind in order to compare it with our more advanced human one? Such a study provided a continuum, then, among various levels of consciousness; the aptitude of the white Norway Rat for this kind of experimentation quickly became apparent, and the maze almost automatically became his learning habitat. (We owe this discovery to W.S. Smart's study of "the mental processes of the rat.")

Thus we have optimum conditions for a comparative psychology, especially a psychology of learning—as long as "consciousness" is our operative term, and as long as "mind" and "consciousness" are our points of reference. The continuum between sub-human consciousness and human consciousness could be compared at various levels, provided one had "consciousness" in the forefront of one's attention.

But what if there were no such thing as "consciousness?" It would then be less useful to *compare* the difference in consciousness if the very subject-matter did not exist. The negative decision about "consciousness" left the door conveniently open for "behaviorism" (the study of overt activity) and then for the "operationalism" which attempted to make the "behaviorist" enterprise completely "scientific." There was an attendant, though minor, inconvenience in comparative psychology. Not many kinds of animals seemed to be able to learn anything (learning-through-discrimination was the pattern), and what they learned was rather limited. To be sure, the training of horses and dogs had gone on for centuries, but university laboratories were not equipped for such activities. The main problem, however, was the consciousness problem. Let us, then, eliminate consciousness, at least from animal psychology, and limit ourselves to the description of animal behavior. If we do so, we need not be so concerned if animals cannot talk to us in our conventional way, to tell us what they are "thinking." We have only to understand what they *do*. This is "objective psychology," and it has a long history which cannot be traced here, although it should be noted that even functional psychology long ago found it more profitable to proceed on objective grounds (how else could "mental tests" have been possible?)

The first great name in "objective psychology," is, of course, Pavlov; and one wonders how much of the subsequent history of American universities and other industries has been conditioned by the kind of experiment which Pavlov undertook—a learning experiment, to be sure, since that was the vogue, but also a learning by conditioning. For it was this kind of objective psychology which Watson noisily espoused in 1913. It was time for scientifically inclined psychologists to rebel against the dogmatic stance of the school of Wundt, with its preference for the study of "consciousness," and its penchant for introspection, not to mention the reams of philosophical writing which issued from it, in spite of its pretension to be a *science*. In addition, neither the "experiments" of the German schools, nor the philosophy which issued from them, seemed coherent. In England, Ryle's criticism of the failures of German introspectionism and of "the ghost in the machine" which German psychology's production seemed to call for, matched (with somewhat more grace) the revolt of Watson. Both charged that Wundt's preference for "consciousness," in spite of his experimental act of faith, was simply the reintroduction into science of the old, outmoded, discredited notion of "soul."

This important step in American psychology, from functionalism to behaviorism, involved, then, a rather ancient stance (one which, in 1845, had even occasioned a formal sworn "pact" against vitalism between von Helmholz and others): the refusal to admit that there are any "other forces than common physical-chemical ones . . . active within the human organism." This pact, and the solemn oath which accompanied it "to establish and compel acceptance of this truth," concretized opposition to the unvoiced conviction among the "new psychologists" that life might involve forces other than those which are discernible in the interaction of inorganic bodies. As we shall later see, the pact was operative not only in psychology, but in biology and in other areas of the human sciences. The principle which Aristotle had named entelechy could no longer be acceptable in scientific psychology or biology. In other words, the classical view that there is an organizing principle which is *other* than the sum of the physical-chemical forces of the living body will no longer be, by common consent, tenable. That a living body and its activities might be considered in terms of totalities (as Dewey later suggested, and as Merleau-Ponty amply showed) was, in a scientific psychology worthy of the name, no longer permitted.

"Comparative psychology," then, dropped out of use as a scientific discipline, only to be introduced later in a reductionist sense—that is, to demonstrate, in our own decade, that men are simply unsuccessful animals—and not, as in its early years, to indicate why men were superior to animals, in spite of all sorts of drawbacks. This reductionism was made possible by two factors: first, that animal psychology had never ceased to be interesting to researchers (rats were in unlimited supply, and all their activities could be perfectly controlled); second, that behaviorism, as a theory about human beings, won the day, and was ready to rejoin its parallel enterprise. Let us see how this happened.

There had been rumblings of discontent in the psychological world for a long time, especially against what seemed a waste of time in translating positively observed behavioral variables in animals into terms of *inferred* consciousness. This was a general malaise; Lloyd Morgan, more specifically, had been complaining about anthropomorphism in interpreting animal behavior. Since, after all, activity—even of animals—is an interesting subject for study in its own right, Watson seemed to have every reason to conclude that the researcher could do without the notion of consciousness. He published his manifesto of 1913, "Psychology as a Behaviorist Sees It," and followed it with the long list of polemical writings for which we know him.

The program was, then, quite simple: One continued to apply the techniques of animal psychology across the board; one retranslated the commonsense notions of "sensation," "perception," "thinking," and so on into terms with which the techniques of animal psychology could best deal: "stimulus and response;" one could even reduce these further to terms qualifying "discrimination," since Pavlov had already showed how one could handle any perceptual problem in terms of the conditioned reflex, understood in terms of measurable discrimination. In fact, it becomes quite simple for one to reduce all "sensory" data to observational operations.

The one chink in Watson's armor was soon welded shut. He had, naively, admitted the validity of the human being's verbal report in psychological investigation; but this smelled of "introspection," and even suggested that human beings might have something to say that could not be measured as an operational variable. Skinner closed that loophole and "verbal behavior" can now be treated as any other form of behavior, measured operationally, and compared (somewhat unfavorably) with the behavior of the pigeon playing a game of ping-pong. "Operant conditioning," as the royal road to "operant behavior," is today the operant behavior of the operant psychologist.

The "scientific" underpinnings of the latest wave in experimental psychology owe much to Bridgman, whose *Logic of Modern Physics* (in 1927) attempted to define scientifically the concept of the scientific concept; for physics, a concept was to mean "nothing more than a set of operations." This means that all concepts are defined by (are synonymous with) the set of operations. This idea was congenial to a science looking for scientific credentials, and for a philosophical, or at least logical, reason for discarding the concept of "consciousness," so long dear to the disciples of Wundt. Since physics, increasingly the model for experimental psychology, could welcome operationalism, psychology could make good use of it. After all, private consciousness must be a meaningless concept for a real science unless it can become public and measurable; but the measurability of a "verbal report" must deal with the subject's having "introspected" about what he is going to report, and that is uncontrollable. The best thing to do is to stick to what can be operationally observed, and forget about the speech of the subject.

Besides, at the very time Bridgman's book was published, the Vienna Circle was saying that, according to the "logic of science" which they were constructing, all scientific language could be and should be reduced to the language of physics. This language rules most of our affective speech out of consideration; thus the "verbal report" usually, in terms understood by Logical Positivism, means nothing. Boring sums up the conjunction well: "The reason for substituting behaviorism for introspectionism in psychology was exactly the reason for substituting operational positivism for experimential positivism in the logic of science." The reason is that we are too much taken up with what scientists consider meaningless questions because there are no means of answering them operationally.

It may seem that the writer, naively "humanistic," feels somewhat threatened by the behaviorist-operationist enterprise. However, if one contemplates the model of humankind offered by this mainline psychology which, after all, reigns in our universities, is practiced by our professors and is transmitted to our students, there is much reason for disquiet.

Carl Rogers, in his well-known debate with B.F. Skinner before the APA in 1956, showed how frightened he was by the rapid strides which had been made toward the understanding, prediction, and control of behavior by the "behavioral sciences." He listed, in his account of their growing power, a few of the frightening aspects of these sciences (which include psychology, psychiatry, sociology, social psychology, anthropology and biology, and sometimes even other social sciences such as economics and political science, with the instrumental disciplines of mathematics and statistics thrown in for good measure). For example, by our use of the behavioral sciences, we are now able "to select individuals who will exhibit certain behaviors; to establish conditions in groups which will lead to various predictable group behaviors; to establish conditions which, in an individual, will lead to specified behavioral results..." What is even more frightening, "in animals our ability to understand, predict, and control goes even further, possibly foreshadowing future steps in relation to man."

This was fifteen years ago, and some of Rogers' worst fears have been realized. In the meantime, David Krech's plea to the APA to consider the *ethical* aspects of "the growing power of the behavioral scientist" does not seem to have registered, as far as the author can determine—quite the contrary. Shortly before his death, in his Presidential Address to the APA (September, 1965) Gordon Allport again warned of the occupational disease of the "galloping empiricism" which has no rational objective, uses no rational method other than the mathematical, reaches no rational conclusion. (The previous week, Pitirim Sorokin had addressed the national convention of The Association of American Sociologists in the same vein.) Allport then saw the tide slowly turning from "itsy bitsy empiricism;" thanks to a growing interest in existentialism in psychotherapy, he thought, a return to consideration of broader issues and philosophical problems was in the offing.

The fears of "humanist" psychologists and their hopes that the tide would turn from experimentation to reflection did not take B.F. Skinner's power into account, however; the prospect of control over human life is much to his taste. In his *Walden Two*, in a sort of science-fiction way which has not failed to attract young students, he has already described the happy utopian society which he hopes will result from the full-scale deployment of operant conditioning. In response to Rogers, he contended that his view of the Good Life *as such* is not opposed by those who believe that man is free, but that they object to anyone's conditioning them into it. There are many absurdities in Skinner's project, not the least of which is the absurd position of the conditioner who will plan the whole scheme and who must necessarily be outside the group of the conditioned, and therefore free. But is the idea of the Master Conditioner only absurd, and not perhaps also horrible? It appears so in Orwell's *1984* (as also, in another key, in Huxley's *Brave New World*).

The ethical problem cannot be solved *scientifically*; it must be solved by *scientists* as ethical beings. The necessity of this is found, for example, in Rogers' sane objection to the project of the scientific control of human behavior. Its greatest flaw is its "denial, misunderstanding, or gross underestimation of goals, or values in their relationship to science." Still debating with Skinner, his counter-thesis was that no matter what scientific endeavor one engages in, whether it be purely theoretical or eminently practical, "there is a prior subjective choice of the purpose or value which that scientific work is perceived as serving," and "this subjective value choice ... must always lie outside of that endeavor and can never become a part of the science involved in that endeavor."

Rogers' point of view agrees with that of Strasser, and illustrates one of the corollaries of the present volume as well:

> *The reality of human freedom can neither be disproved nor proved by empirical methods, for this freedom is the basis of all human activities,* including that of pursuing science... Even though *empirical* human science works with exact, statistical methods, it *does not imply a deterministic philosophy* ... Man's freedom is presupposed in all the human sciences... consequently it cannot be a term of this discourse. In other words, freedom is not a 'factor' in the empirical sense...[5]

Why does the statistical method of empirical human science not imply a deterministic philosophy? In the field of sociology, for example, where the relationships of greater numbers of human beings are considered, the answer can perhaps be seen more clearly, and Gurvitch has put it well:

[5] Stephan Strasser, *Phenomenology and the Human Sciences,* Pittsburgh, 1963, 285.

> Sociology is the science of human freedom and of all the obstacles which this freedom encounters and overcomes in part. The other sciences, whether they be called economics, law, moral sciences, anthropology, human geography, or demography, etc., are distinct from sociology only through the limitations of the direction taken by the effort (i.e. to overcome the obstacles) and by the selection of the obstacles to be overcome. The reality studied by all these sciences is the same—namely, the human condition, viewed in the particular light thrown by the scientific method. [6]

Radically, the answer is that all these sciences deal with the limited freedom of the human individual. When one speaks of the "human condition," as Gurvitch does, one always intends that network of human liberties of which Merleau-Ponty wrote, in which the freedom of each is limited by the freedom of all the others (these are some of the "obstacles"), and each of these liberties is, in itself, radically finite. It will not be surprising then, that the human situation offers only a limited number of possibilities for human action, and that consequently one can plot the successes and failures to overcome obstacles on a statistical curve, even though the curve will never reveal which are the successes and which the failures, unless they are designated as such before the curve is plotted.

These issues of human liberty, morality, and determinism do not seem out of place here, even though the spokesmen and the issues spoken about appear in other contexts in this book.

If the reader will re-trace the steps in the enterprise of experimental psychology, he will find, as suggested in the earlier pages of this chapter, the same anti-humanist bench-marks running through the psychological model of man. Ernest Becker thinks that Watson's original endeavor was "hopeful" in spite of its being "incredibly narrow;" but Skinnerian psychology is even narrower, is it not?—and willingly so, in its attempt at precision (what Koch called its "scholasticism").

To be sure, Watson's incredibly narrow behavioristic psychology *was* hopeful, provided it did not pretend to the all-inclusiveness of its model. But it *has*, perhaps contrary to the will of most of its practitioners, become an all-inclusive model, aping the universality of the philosophical doctrines from which it tried to free itself in becoming a science. Oddly enough, it is this blindness to philosophical issues, I think, which has most limited the psychological enterprise. Boring, whose extremely valuable and ency-

[6] Georges Gurvitch, "Réflexions sur les rapports entre philosophie et sociologie," *Cahiers Internationaux de Sociologie*, 1957, (22), 10. (Translated by Strasser, p. 285.) (Unless otherwise noted all translations are by the writer.)

clopedic *History of Experimental Psychology* makes every effort to show us psychology as a truly scientific enterprise, is one of the psychologists who scorns philosophy most; yet, as Becker points out, Boring's final choice for the greatest father figures of psychology, Darwin and Freud, shows by his very choice that "psychology has no business being separated from philosophy." [7] I agree with Becker that "It has been a relatively fruitless journey, and has exhausted many talents, this anti-philosophical disciplinary quest of psychology," yet I do not share his view that psychology is "beginning to show some signs of emerging from its long scientific sleep and a willingness to pick up again the current which the discipline abandoned at the end of the nineteenth century..." [8]

There is, first of all, a question of prestige. Now that psychology has become a "science" in its own right, with at least a technique to show for itself, can the psychologist turn to philosophy to "fill in the gaps?" I have maintained above that this is impossible, for it is "unscientific." Nor are any philosophical reflections upon the *results* of experimentation in animal and human psychology of any use unless they can be integrated into a second order reflection. That is, a reflection which explicitates the "pre-understanding" of self, other, and world which the sciences further explicitate in various ways. Those psychologists who are still my friends assure me that, after all, it is better to play the game, to run the rats, to make the measurements, and to forget about the possible philosophical implications or contradictions. But this brings up a second question: that of *money*. Let us face it: we are all aware now of the "one-ahead" method of psychology.[9] This consists in the experimenter's designing and completing his experiment with his own resources; then, since requests for research funds must be accompanied by detailed explanations of methodology, projections about results, and estimates of cost, he is in a good position to present an excellent research project. When such an impressive plan is awarded equipment and research help,

[7] Ernest Becker, *The Structure of Evil: An Essay on the Unification of the Science of Man* (New York, 1968), 116. Becker's argument: "After all, it was the theory of evolution which conclusively introduced the idea of process into the organic realm, thereby attacking the primacy of static measurement of regularities in the human sciences. Has experimental psychology treated man in an open-ended way, as a creature in a continual process of becoming? As for Freud... it is really difficult to understand how a careful laboratory scientist could be comfortable with the scientific shoddiness of Freud's subjective method, much less with Freud's avowal that psychoanalysis has its own distinctive mythology and rhetoric. Have experimental psychologists really tried Freud's type of science? Even worse, Freud helped to undermine the very disciplinary integrity that Boring so eagerly covets..." See also, on this latter point, Paul Ricoeur, *Freud and Philosophy* (translated by Dennis Savage), New York, 1970.

[8] *Ibid.*, 117.

[9] David Bakan, in his book, *On Method,* revealed this method to the non-scientific public.

the first project can be published and the money, equipment, and assistantships used to prepare the next assault upon a university or national purse, the keepers of which have become collectively dazzled by the magic of "science," by hope in "scientific progress," and by the promised "betterment of all mankind" through the onward march of "science." Third, there is the problem of *ignorance*. One can teach a technique to a student, and he can become a "scientist" upon the paradigm-worker model to which we have now unfortunately become accustomed; one cannot teach him as easily to reflect upon what he is doing—especially if such thought is unprofitable. Even Einstein asked a philosopher not to require of him to say *how* he thought, but to examine what he *did*. Even excellent scientists carry about with them the formulations of a third-rate epistemology; and it is not at all unusual for them to object to the philosopher's attempt to reflect upon what they are *doing*.

This is a not-too-polite way of saying that psychology is too important to be left solely to the psychologists. Experimentalists bring to the task of elaborating their theories *a priori's* which are not capable of critical elaboration within the context of the experimental situations which give rise to them. With regard to certain problems, elaborate theorizing is often done upon insufficient observational basis, with the result that the theory appears incomplete. It moreover gives the impression of philosophizing not in terms of the observed data but in virtue of the *a priori's* which motivate the enterprise. The overall impression is, then, that of a vague sort of anthropology without man. The perfectly respectable custom of reinterpreting the "classical" experiments in the contexts of new theories represents an honorable attempt to bring within the confines of wider theories an increasing number of empirical facts; but that this procedure is at all possible invites reflection. However, that data collected in the context of one theory may find place in a theory diametrically opposed to it need not lead to the relativist conclusion that they both must be, in a sense, *wrong*. It may also lead to the phenomenological conclusion that they may both be, in a sense, *right*. The task is, then, to discover in what fundamental intentionality they may both find meaning. Erwin Straus has put the matter succinctly: "As a matter of fact, objective psychology cannot exist without a black market furnished with contraband from the psychology of living experience." [10] This is a strong statement but the theme is recurrent. Köhler, in order to find the place

[10] Erwin Straus, *The Primary World of Senses: A Vindication of Sensory Experience*, trans. Jacob Needleman (Glencoe, 1963). (Straus is here speaking of the world of experiencing which is the taken-for-granted starting point of all sciences.)

of value in a world of facts, discovered that "description" was not enough, and that it was further necessary to call upon phenomenology as "the field in which all concepts find their final justification." [11]

The "anti-philosophical quest of psychology" has led us to a concept of man which in its exclusiveness tends toward robotization; it is the result of making a philosophical conclusion on the basis of an anti-philosophical technique. However, it must be admitted that the psychological technician had little philosophical help in his efforts, and little to fight against in the camp of philosophy; and it seems no wonder that the results have been disappointing. The basic disappointment may be restated: Behaviorism, and *a fortiori* operationalism, limits its activity to the observation of overt activities and reactions, and its interest is exclusively in the quantitative nature of such activities and reactions. Its basic defect has been its constant attempt to eliminate the subject from psychology, "as if," remarks Georges Thinès, "the subjective could not achieve status in a scientific vision of man except insofar as it is reduced to physical or physiological dimensions." [12] Its limitation of the field of research has been arbitrary, its generalizations too hasty. In a larger context, a criticism of behaviorism-operationalism may be summarized in criticizing its dependence upon a mechanistic physics. It thus presents the obverse of the traditional "spiritualist" psychology while sharing its methodological defect, namely, that of shaping its evidence to suit a presupposition.

It is not entirely justifiable, however, to speak of the "traditional spiritualistic psychology" as a homogeneous realm of thought. There remain diverse tendencies within the contemporary movements of philosophical psychologies with a spiritualist commitment.

One of these tendencies (partly represented by Donceel) seems to contend that a purely metaphysical psychology is no longer possible, and that the psychology of human beings is inconceivable without the implication of the total human organism. Consequently, psychology must be considered a part of the philosophy of nature. This approach, therefore, demands that valid reflection upon psychological events must take its departure from experience. The type of experience envisaged, however, is not that of simple subjectivist introspection; in keeping with the realist tradition it follows, this tendency envisages experience as that of the objective observation of man as an organism integrated within the cosmos.

[11] Wolfgang Köhler, *The Place of Value in a World of Facts* (New York, 1938), 102.

[12] Georges Thinès, "Horizons et impasses de la psychologie actuelle," *Revue philosophique de Louvain*, 1961 (59), 94.

The results of the positive sciences are thus welcomed, and the task of philosophical psychology is then the discovery of the deeper significance of the results achieved by the positive sciences. Simple observation and description, it is here justly felt, deliver knowledge only of matters of fact. Reflection will thus render traditional philosophical concepts more meaningful and precise, while giving to empirical facts their full significance. In this program of "despoiling the Egyptians," the anti-metaphysical mechanistic scientists are supposed, it seems, to get their comeuppance.

However, such a program is illusory. "Matters of fact" will never, of themselves, reveal their "how" and "why," and philosophy becomes a matter of interpretation of such "facts" on the basis of a metaphysical theory. The interpretation of matters of fact thus furnished by careful scientific observation is carried out within the context of a hylemorphic cosmology which is already encompassed within the context of a fully systematized metaphysics, in which everything-in-general is already explained. The results of modern experimental sciences fit very badly into the matter-form, substance-accident context of this cosmology; so reports of experimental research must be broken up and recast ever so eclectically into categories and terminologies more acceptable and more apt to render unto them their true "significance." This criticism is a reiteration of the plea for the kind of "presuppositions" to be defended in the conclusion of this book as the necessary givens in the starting-point of philosophy, and a cry of astonishment that Donceel can, after what he calls a "phenomenological" study (which "has shown that we live in the midst of objects or quasi-objects") insist that "these quasi-objects can exist only for a subject." This is in itself not a wholly shocking statement until the "subject" is defined: "an originating, primordial I, this ego-source, is *my soul*, is I as *soul*."[13] The point here is not to refute Donceel, but to reassert that the philosophical critique attempted in this book belongs in the tradition of those who have attempted to study man without the presuppositions of *either* Immanence *or* Mechanism, and to do so even beyond the presuppositions of a subject-object dichotomy. I do not quarrel with the concept of "soul," let it be noted, but with the easy equation of "subject" with "soul" in the context of a cosmological investigation! There is always the danger, in this kind of approach, that experimental evidence will be sifted to provide those elements which can be integrated into an acceptable metaphysical context. At best, this

[13] Joseph Donceel, *Philosophical Anthropology*, New York, 1967, 410.

method is simply another instance of the reinterpretation of classical experiments to fit a special theoretical context. It is at worst incoherent. Might one also generalize the reproach of Paul Ricoeur against the "cosmology of liberty" into a reproach of a "cosmology of the psyche" in general? [14]

Another tendency in philosophical psychology is represented by those who would base their reflections not upon established experimental, or empirical, evidence, but upon "everyday" experience, in the grand Aristotelian manner. Donceel comes to it in a penultimate chapter; but one thinks first of Maréchal. In contrast to the Aristotelian manner, however, the "everyday" experience is not that of empirical observation, but of subjective introspection: upon ordinary speech, upon insight, or more fundamentally upon the pre-scientific and originating experience of the self. In further contrast to the Aristotelian manner, it constructs its metaphysics not through a prior elaboration of a cosmology, but directly, taking a short cut to its goal and avoiding discussion of positive scientific results as having been artificially chosen and developed (as indeed they have been). There is thus no need to presuppose the validity of scientific findings; the experiences chosen for consideration are pre-scientific, concrete, and concretely experienced (though "explained" in metaphysical categories). The chief task of philosophical psychologists who favor this transcendental approach has been to produce a metaphysical psychology of great interest and power; their main objective has been to establish the "conditions of possibility" of the ordinary experiences of human life.

In this context, Karl Rahner deserves very special attention. Now that his basic works, *Geist in Welt* and *Hörer des Wortes* are available in English translation, together with other important theological writings, his extremely original contribution can receive the attention and discussion it deserves upon these shores. The rich and masterful study by Andrew Tallon will be of inestimable help to readers.[15]

[14] See Paul Ricoeur, *La philosophie de la volonté*, Vol. I: *Le volontaire et l'involontaire* (Paris, 1949), 172-186. In the section on "determination and indetermination," Ricoeur defends his option to seek the living experience of the subject in the subject himself with regard to his self-determination rather than to attempt an *a priori* Thomist demonstration of the freedom of indetermination. One of his basic reasons is that a consideration of will-as-nature and as part of a philosophy of nature introduces an "element of necessity" into the will. The whole argument can be read in Erazim V. Koháks's translation (*Freedom and Nature: The Voluntary and the Involuntary*, Evanston, 1966).

[15] Andrew Tallon, *Person as Personization in Karl Rahner's Philosophical Anthropology*. We can only hope for a very early publication of this work. The writer has had access to Mr. Tallon's manuscript, and thus does not feel permitted to reveal the substance of the book. However, a brief remark about its intention will not be out of place as giving rise to a few remarks of the writer's own.

Rahner's intention, in the tradition of Maréchal, has been to take his point of departure for metaphysics in an examination of knowing. Tallon remarks that for Rahner the examination of knowing is not undertaken for its own espistemological sake, but in view of an ontology of man. Tallon's own interest in Rahner's work has not been precisely in the metaphysical aspects of being as derived from and applied to knowledge. His intent is rather to apply that metaphysics to the larger question of becoming—how spirit becomes itself through matter.

The reader will expect that, in the Maréchallian tradition, personization (the dynamic and open-ended relationship of *becoming* between the constituent principles of the human person, Spirit and Matter) will necessarily be based upon cognition. Without forestalling Tallon's effort, and understanding personization within the general context of Thomist-Maréchalian-Rahnerian philosophical anthropology, it would indicate that person is both spirit and matter, that the incarnation of spirit is the source of freedom as of cognition (that is, of will as of intellect), that judgment is the sign of perfection of both cognition and freedom. Whatever be the dialectical processes of the polarities of spirit and matter, writers in this tradition give primacy to spirit as the essence of human being, and equate intellect with spirit. (Donceel says "soul.") However, it should be pointed out that, with all their caution, these writers, including Rahner, will have an impossible task in keeping spirit incarnated, even though there is no hint of Manicheism in their intention.[16]

Self-becoming through ethical acts is an attractive and realistic thesis, and Tallon well emphasizes a break in Rahner's work with an old-time atomization of behavior prevalent in classical ethics and moral theology, and a positive tendency toward a holistic approach which rejoins the efforts of, say, Louis Monden. But without a phenomenology of the body, ethical acts continue to be appraised on a cognitive level; and why conscience does not, then, enjoy an infallible rectitude will be a primordial and unanswerable problem. The half of Donceel which is Rahnerian and Maréchalian, we have seen, has drawn this ontological conclusion with little insight but with inexorable logic.

[16] In William Dych's translation of *Geist in Welt (Spirit in the World,* New York 1968) we read (p. 248): "We must show that this line of thought is Thomistic." Why we must do so is not clear; but a quick check through Rahner's impressive list of backup references to Thomas shows that Rahner is not using the same set of texts as Ricoeur had used. The exceptions are *S.T.*I., 82 (which Rahner uses only as an example of a "general principle") and *S.T.* I-II, 13. Rahner's base text seems to be *S.T.* I, 77, and he avoids question six of De Malo; Ricoeur sees in the texts *he* chooses as representing Thomism "subordination of indeterminism regarding the finite to determinism regarding the infinite"—which I believe is Rahner's intention.

There is much to recommend this general approach: its immediacy, and its reliance upon lived experience, in particular, render it especially attractive to contemporary thinkers. However, when second thoughts arise, one cannot help inquiring about just what it is that is experienced. There seems to be involved the old myth of introspectionism, the myth that what we are conscious of is so immediately apparent to us that we may consider it as simply given and pass on to reflection on the "conditions of possibility" of our consciousness.[17] Part of this myth seems to require the Cartesian commitment to belief in the infallibility of immediate consciousness. Once this is called into queston as the starting-point for reflection, all certitude is rendered impossible, and the project justly merits the criticism of Ryle noted above, in his condemnation of "the ghost in the machine."

The point at issue is that in this endeavor, the interest of philosophers in the "conditions of possibility" of a given originary experience far overshadows the importance of the experience itself. One is tempted to resort too soon to the exploration of a transcendental ego as that condition of possibility, and to do this before one has taken the pains to investigate the structure, the mode of appearance, and the constitution in consciousness of the phenomenon in question. In other words, this approach constitutes something very like a return to immanentism (or at least subjectivism) in philosophy. At least it shares with both philosophical and anti-philosophical psychologies that sharp all-or-nothing option which seriously compartmentalizes and restricts attention, and which cuts large areas of fruitful reflection out of consideration. It surely overlooks the importance of a disciplined subjectivity.

Are we, then, in the "double-bind," "no-win" kind of situation? Neither the search for the conditions of possibility of an ill-defined experience nor the dependence upon a purely descriptive approach to experiencing will suffice. But what then? Is the necessary corrective to the first position to be found in a descriptive method, and the corrective to the second to be found in prior philosophical understanding? The answer is: In a sense, yes. Description without understanding is meaningless; understanding without description is empty.

But what kind of "understanding" is involved? Understanding of a previously organized metaphysics is one way, of course, and Rahner has

[17] In what is probably the key chapter in Rahner's *Spirit in the World,* one reads: "... it follows that in this chapter our task cannot be that of proving that the conversion to the phantasm is a factual characteristic of human knowledge.... Rather what we must do is grasp it in its possibility." (238)

proceeded upon these lines; but this reader can report the oft-repeated experience of students, unprejudiced and intelligent, confronted with the grand metaphysical program of Thomas Aquinas, refashioned for the classroom by, for example, Farley and McGlynn or Owens. The immediate reaction is their impression of a certain "quantification of perfection," and more basically, the inability to find any real or theoretical distinction between *persons* and *things* in the system. Johannes Metz has recognized the insufficiencies in the Thomistic world of objects, and prefers to find the world as one of persons. But if person is spirit-matter (even in the extremely careful analysis of Rahner), and if *spirit* inexorably becomes *person*, then it seems time to begin from the other end, with a phenomenology of the body and of the *givenness* of person, understood primordially in the lived world of experiencing.

Thus, a "hermeneutic spiral" does not cease to weigh upon the reflections of these chapters and to direct the effort at critique.[18] The static, meaningless descriptive character of experimental psychology has been found inadequate to explain human being in the world; but a transcendental reflection without adequate empirical bases is also unsatisfactory. The distance between the knower and the known cannot be bridged, much less encompassed by either approach. A new effort must be made.

What is needed, then, is a method which, first of all, takes into account both the fruitfulness and defects of the methodological endeavors heretofore discussed, but which at the same time, by its own coherence, avoids a meaningless eclecticism or simple concordism (e.g. the effort in Catholic seminaries to correlate "the two psychologies"). What is needed is a method which refuses the pure subjectivism which has voided so much of the effort at introspection, avoids the purely mechanical reporting of the quantitative observations of overt behaviors (which cannot, as simple descriptions of activities in space-time categories, mean anything), avoids forcing objectively reported matters-of-fact into prearranged metaphysical categories, and avoids the oversimplified hypothesizing which regards a "transcendental ego" (thought of as "pure subjectivity") as the necessarily *a priori* condition of possibility of human experience.

On the positive side, what is needed is a method which can combine a careful description of pre-scientific, pre-philosophical experiencing with a valid ontological reflection (that is, a reflection which has a metaphysical intention but which proceeds with full respect for the structure of the

[18] See chapter VII.

phenomenon experienced and for the structure of the experience in which it is given). Such a method must give assurance that an experience has actually taken place, and must offer the hope that its delineation will aid in the discernment of the structure of that experience. It must be, in short, a method which can provide an objective report of individual, pre-scientific, pre-philosophical subjective experiencing. Such a method is, assuredly not an easy order; and where can one find an assemblage of valid experience-reports of human subjectivity?

The effort has *not* been made, as we have seen, by experimental psychology. It was made rather by those involved in what is called "third force" psychology: the group whose work clusters around the names of Allport, Maslow, Rogers, and their antecedents. ("Third force" psychology is thus distinguished from the first, experimental, psychology, and from the second, or Freudian, psychology.) We must turn attention to these movements in later chapters.

SELECTED INTRODUCTORY READINGS

Adler, Mortimer J., *The Difference of Man and The Difference it Makes, N.Y.*, 1967.

Arendt, Hannah, *The Human Condition: A Study of The Central Dilemmas Facing Modern Man,* N.Y., 1959.

Barfield, Owen, *Speaker's Meaning,* Middletown, Conn., 1967.

Becker, Ernest, *The Structure of Evil: An Essay on The Unification of The Science of Man,* N.Y., 1968.

Bronowski, J., *The Identity of Man,* New York, 1966.

Chisholm, Roderick M., ed., *Realism and The Background of Phenomenology,* 2nd. ed., 1967.

Cirne-Lima, Carlos, *Personal Faith,* New York, 1965.

Coulson, William R., and Carl R. Rogers, eds., *Man and The Science of Man,* Columbus, 1968.

Doniger, Simon, ed., *The Nature of Man in Theological and Psychological Perspective,* New York, 1962.

Edie, James M., ed., *Phenomenology in America: Studies in The Philosophy of Experience,* Chicago, 1967.

—, *New Essays in Phenomenology: Studies in The Philosophy of Experience,* Chicago, 1969.

Farber, Marvin, ed., *Philosophic Thought in France and The United States,* New York, 1968.

Gurwitsch, Aron, *Studies in Phenomenology and Psychology,* Evanston, Ill., 1966.

Hart, Ray L., *Unfinished Man and The Imagination: Toward an Ontology and Rhetoric of Revelation,* New York, 1968.

Hems, John M., "Husserl and/or Wittgenstein," *International Philosophical Quarterly,* 1968 (8), 547-578.

Husserl, Edmund, *The Crisis of European Sciences and Transcendental Phenomenology,* trsl. and introd. by David Carr, Evanston, Ill., 1970.
Ihde, Don, "Some Parallels Between Analysis and Phenomenology", *Philosophy and Phenomenological Research,* 1966 (27), 577-586.
Jordan, Robert Welsh, "Husserl's Phenomenology as a 'Historical' Science," *Social Research,* 1968 (35), 245-259.
Kockelmans, Joseph J., *A First Introduction to Husserl's Phenomenology,* Pittsburgh, 1967.
—, ed., *Phenomenology: The Philosophy of Edmund Husserl and its Interpretation,* New York, 1967.
Koestenbaum, Peter, "Existential Psychiatry, Logical Positivism, and Phenomenology," *Journal of Existential Psychiatry,* 1961 (1), 399-425.
Laing, Ronald D., *The Divided Self; An Existential Study in Sanity and Madness,* London, 1960.
—. *The Self and Others,* London, 1961.
—. *The Politics of Experience,* New York, 1967.
Montagu, Ashley, *On Being Human,* London, 1957.
Platt, John R., ed., *New Views of The Nature of Man,* Chicago, 1965.
Polanyi, Michael, *The Study of Man,* Chicago, 1959.
—. *Personal Knowledge: Towards a Post-Critical Philosophy,* 2nd. ed., New York, 1964.
—. *The Tacit Dimension,* New York, 1966.
Ricoeur, Paul, *Husserl: An Analysis of His Phenomenology,* trsl. Edward G. Ballard and Lester E. Embree, Evanston, Ill., 1967.
Shinn, Roger L., *New Dimensions in Theology Today,* vol. VI. *Man: The New Humanism,* Philadelphia, 1968.
Spiegelberg, Herbert, *The Phenomenological Movement: A Historical Introduction,* 2nd. ed., The Hague, 1965.
Strasser, Stephen, *Phenomenology and The Human Sciences: A Contribution to a New Scientific Ideal,* Pittsburgh, 1963.
—. *The Idea of Dialogal Phenomenology,* Pittsburgh, 1969.
Tillman, Frank A., "Philosophical Analysis and Phenomenology," *International Philosophical Quarterly,* 1966 (6), 465-482.
Zaner, Richard, "An Approach to Philosophical Anthropology," *Philosophy and Phenomenological Research,* 1966 (27), 55-68.

SELECTED ADDITIONAL READINGS

FOR CHAPTER I

Bakan, David, *On Method: Toward a Reconstruction of Psychological Investigation,* San Francisco, 1969.
Boring, Edwin G., *A History of Experimental Psychology,* 2nd. ed., New York, 1957.
Bridgman, Percy William, *The Logic of Modern Physics,* New York, 1927.

Dewey, John, "The Reflex Arc Concept in Psychology," *The Psychological Review,* 1896 (3), 357-370.

English, Horace B., and Eva Champney English, *A Comprehensive Dictionary of Psychological and Psychoanalytical Terms: A Guide to Usage,* London, 1958.

Eysenck, Hans Jurgen, ed., *Behavior Therapy and The Neuroses: Readings in Modern Methods of Treatment Derived from Learning Theory,* London, 1960.

— and S. Rachman, *The Causes and Cures of Neurosis : An Introduction to Modern Behavior Therapy Based on Learning Theory and The Principle of Conditioning,* London, 1965.

Hebb, Donald O., *The Organization of Behavior: A Neuropsychological Theory,* 2nd. ed., New York, 1961.

Heidbreder, Edna, *Seven Psychologies,* 2nd. ed., New York 1961.

Ryle, Gilbert, *The Concept of Mind,* 2nd ed., New York, 1961.

Skinner, B.F., *Science and Human Behavior,* New York, 1953.

—. *Verbal Behavior,* New York, 1957.

—. "Behaviorism at Fifty," *Science,* 1963 (140), 951-958. (Also in *Behaviorism and Phenomenology,* ed., T.W. Wann, 79-97.)

Wann, T.W., ed., *Behaviorism and Phenomenology: Contrasting Bases for Modern Psychology,* ("The Rice Symposium") Chicago, 1964.

Watson, John B., "Psychology as a Behaviorist Sees It," *The Psychological Review,* 1913 (20), 158-177.

—. *Psychology from the Standpoint of a Behaviorist,* Philadelphia, 1929.

CHAPTER II

THE HOLISTIC MODEL

COMING CLOSE TO THE TOTAL MAN

We have seen in the preceding chapter that the warring factions of the disciples of Wundt created two sorts of revolt: the first, in the form of functionalism, respected consciousness but rejected the "elementism" of the master; the second, in the form of behaviorism, kept the ideal of elementism and eliminated the idea of "consciousness." It was the second revolution which won out in mainline experimental psychology in America, while the functionalist revolution made some headway in the social sciences, but lost out, generally speaking, in the control of the enterprise of experimental psychology. The operationalist takeover was further simplified by the fact that functionalism as a technique could be incorporated into the behaviorist enterprise without incoherence. In the endeavors of *philosophical* anthropology, the empiricist successors of Aristotle seem increasingly threatened with irrelevance, while the transcendental philosophers seem incapable of maintaining a link with empiricism which would make their endeavors more truly a philosophical *anthropology* by integrating somehow the human body into their considerations. There is, it seems, more than a hint of dualism in the latter effort. Perhaps a holistic model of human being in the world may contribute a larger context into which transcendental philosophy might be at home. Let us see. In providing such a model, the present thesis maintains that the holistic model belongs in a phenomenological tradition.

Let us not lose sight of the fact that in the beginning, the conflict was not between Wundt and his successors; they fought out their battle independently of him. The original option, as Boring pointed out, was between Wundt and Brentano, between experimentation and empiricism, between the preeminence of sensory *contents* and that of psychic *acts*. Consequently, something must be said about the pilgrimage of act psychology and its admittedly minor, though hopeful, presence on the contemporary scene.

Both movements (let us call them behaviorist and phenomenologist) had a common source in the attempt to gain recognition of the new psychology of the nineteenth century as a science, even though both Wundt and Brentano were philosophers. However, the phenomenological tradition from Brentano through Stumpf and Husserl to contemporary third force psychologies is a different kind of opposition to the school of Wundt by its descendents of every stripe. Brentano's *Psychology from an Empirical Standpoint* appeared, as if by chance, the same year as the first volume of Wundt's *Beiträge.*[1] Brentano's work was directly connected with perception, and this will be the theme of all the movements which owe their origin to him. In the transcendental phenomenology of Husserl, for example, as in the perceptual experiments of the Gestalt psychologists and philosophical reflections of Merleau-Ponty, the model is that of *vision*—and, by analogy, *intuition.* Brentano's call for the reformation of philosophical method was to be implemented by an adoption of more rigorous methods, modelled upon the empirical methods of the natural sciences. Brentano's *Habilitation* at Würzburg, defending this proposition, was of capital importance in the careers of many young philosopher-psychologists, and chief among them was Carl Stumpf.

Before anything can be made of Stumpf's relevance to contemporary psychology and to our study, the importance of Brentano should be made clearer. Tichener, psychology's great "displaced person," was categorical in his assertion that "there is no middle way between Brentano and Wundt," [2] and that Brentano would never have been an experimental psychologist even if he had had a laboratory. However, Rancurello, Brentano's contemporary translator, biographer, historian, and bibliographer, maintains that Brentano's experimentation would simply have been different from Wundt's; and the present writer would suggest that it would probably have taken the form of Stumpf's empirical studies. Perhaps these reflections are neither here nor there; what is important is that Brentano, though unable by force of circumstances to found an experimental laboratory, was convinced of the necessity of an experimental foundation for both act and content psychology.

Brentano's basic project was the construction of a descriptive (he called it also "phenomenological") psychology as a complement to the

[1] It was remarked at the beginning of chapter I that the "philosophical anthropology" in view here had its origin in Germany in 1874. It may be helpful to remark that third force psychology began that same year, with Brentano.

[2] E.B. Tichener, "Brentano and Wundt: Empirical and Experimental Psychology", *The American Journal of Psychology,* 1921 (32), 108-120.

explanatory (he said also "genetic") psychology of the school of Wundt. His opposition was directed mainly toward the *exclusiveness* of Wundtian explanatory procedures; Brentano was convinced that descriptive methods uncover the theoretical foundations of all sciences, and that explanatory theories, coordinated with descriptive procedures, draw their meaningfulness and relevance from description. Description was to be directed toward the "psychic act," or rather the "psychically active subject." The descriptive methodology was to be a study of the "inner perceptions" of the subject as the key to the sources of all knowledge. Rancurello observes:

In essence, Brentano was convinced that descriptive psychology was necessary to 'gain an insight into the true origin' of every type of human knowledge, be it in the realm of metaphysics, logic, natural sciences, or *Geisteswissenschaften*. In particular... he was convinced that not only the specific problems of aesthetics, logic, epistemology, and ethics, but the problem of human existence as such, could not be adequately solved without first investigating its psychological foundations; this conviction prompted the original plan of his psychology.[3]

Correlated with descriptive methods was to be the observation of all the ways in which these acts are externalized, in opposition to his contemporaries' reduction of psychology to physiology (and, by simplification, rather the *contents* than *acts* of "psychic activity") and thus their concentration upon *structure*. Rancurello argues that such "contents" are "twice dead," and that structuralism leads neither to true explanation nor understanding, since it cannot take the dynamic character of the psychically active subject into account.

Thus, Brentano's effort was directed mainly against the atomism, sensationism, associationism, and elementism of the school of Wundt, and toward a holistic attitude which would lead later to the "free study of phenomenal wholes" undertaken by the Gestalt psychologists.[4]

It is evident that our intention here is to link Brentano with the holistic

[3] Antos C. Rancurello, *A Study of Franz Brentano*, New York and London, 1968, 15. Rancurello's study of Brentano serves largely as the basis of these remarks on the place of Brentano in the formation of an "empirical phenomenology" as the basis of the holistic model of man.

[4] For a good summary, see Rancurello, 80; and for purposes of clarity and relevance to the present study, consider the following: Since Brentano thought his descriptive or phenomenological method primordial, his view of explanatory or genetic psychology as a correlate to descriptive method is explained: "Brentano's genetic or explanatory psychology corresponds to the traditional model of this science. As used by our author, the term 'genetic' indicates that this fundamental branch of psychological enquiry investigates the 'laws' governing the 'genesis' of psychological processes or states and their 'externalization' in behavior; the term 'explanatory' adds the important connotation that the discovery of these laws satisfies all the requirements of a true explanation . . ." (39)

views of contemporary third force psychologies. In order to do so, we must consider the link provided by Stumpf and the Gestalt psychologists.[5]

It may seem surprising to find the name of Stumpf at the head of the list of contemporary third force psychologies. Though Stumpf died only in 1936, not so long ago in terms of years, his name seems to belong to a bygone age in psychology. Between him and contemporary researchers in the behavioral sciences, there has intervened a whole series of revolutions—not only in psychology, but also in sociology, art, politics, religion, and popular culture. Furthermore, Stumpf's *Selbsdarstellung*, a review of his life and work, appeared as early as 1924, when he was already 75 years old. However, the current of thought which owes some of its beginnings to him, and which may have contributed precisely to the series of revolutions which succeeded him, has survived. In spite of the sea changes undergone by all currents of thought, Stumpf's contribution to the psychology of perception and to phenomenology warrant our considering him a contemporary.

Stumpf, as a young student, was present at Brentano's *Habilitation* in Würzburg in 1866, and it was of capital importance for him. From that date on, he pursued his philosophical investigations in an "empirical" manner, rejecting the notion of knowledge as "looking at," but deriving it nevertheless from an analysis of empirical data. His method was phenomenological; he espoused phenomenology in the sense of a prescience, an initial descriptive analysis of the perception of phenomena (especially sound and color). In short, his was an empirico-experimental phenomenology, with a methodology akin to that of Husserl, as Spiegelberg points out.[6]

In this connection, then, it is important to note that Stumpf's empirical studies in perception were undertaken, even at that early date, in a context of philosophical reflection. Boring, of course, is not happy with this, since such a procedure violates the canons of positivist "scientific" psychology; but it means that the movement which took its inspiration from Stumpf was going to be less likely to succumb to a purely statistical view of psychological research, and more apt to be open to philosophical development, though its methodology was to be chiefly empirical. Stumpf's

[5] Rancurello (98-134) makes the connection more tellingly, taking into account the complexities of the question. The present account must be extremely simplified, and directed toward the establishment of our own thesis.

[6] Herbert Spiegelberg, *The Phenomenological Movement*, 2nd ed., The Hague, 1965, 53. See also Boring 362-371.

own philosophical situation is interesting in itself; his predecessors in empiricism were Locke and Leibniz, as well as Brentano with his Aristotelian background. He was familiar with, and appreciated, the work of the young John Dewey; he carried on a lifelong correspondence with William James, and praised James' work as "the best of all psychologies."

Stumpf's work deserves mention here because he shared with Husserl the influence of Brentano and a phenomenological method. If being a student of Brentano is an argument for one's being, at heart, a phenomenologist, Stumpf deserves the name far more than does Freud. If Freud's one semester at the feet of Brentano provides a minor reason for speaking of him in a phenomenological context, how much more right has Stumpf, who shared with Husserl the honor of contributing *Erinnerungen an Franz Brentano* to Kraus' biography of their master in 1919. Even at that date Stumpf and Husserl were considered the foremost of Brentano's students, each with a special relationship to Brentano, although only Kraus was left to bring painfully to light the unpublished works of the tragic career of Brentano. It should not be forgotten either that Husserl was also a student of Stumpf, prepared his *Habilitationsschrift* under Stumpf's direction, and dedicated his *Logische Untersuchungen* to Stumpf "in honor and friendship." If it is possible to reconcile the two divergent trends within phenomenology itself—the experimental and the transcendental—which parted company at the very beginning of the phenomenological movement, (this reconciliation is intended here) the reconciliation must take place with the help of Stumpf as well as that of Husserl.[7]

Stumpf's attachment to phenomenology, in the special sense proposed here, is a quite plausible thesis. His influence upon subsequent events is also defensible.

Before continuing with an account of Stumpf's relevance to contemporary psychology, a few points need to be made clear, especially

[7] There has been a good deal of "muttering in the ranks" in recent years over the apparent fact that *Husserl* has become the primary phenomenon in phenomenological investigations. Some phenomenologists continue to be intensely interested in Husserl's "transcendental" phenomenology, while others claiming the title "existential phenomenologists" are more anxious to "get on with the job" of studying the empirical ego and the *Lebenswelt.* It is the author's conviction that the latter task cannot be meaningfully accomplished unless the "transcendental reduction" is recognized. The split in phenomenological focus, it seems, is due to Fink, not Husserl; it was Brentano who correlated "description" and "experiment" most successfully, and Stumpf who carried on to found a tradition which the author pleases to call an "empirical phenomenology" which remains faithful to the intentions of Brentano and Husserl—and indeed is possible only because of them.

the relationship between Husserl and Stumpf, and between them and Brentano.[8]

Brentano's chief insight was that of "intentionality," that is, that every consciousness is "consciousness *of*," directedness toward, relationship to. This, for Husserl, was as much a manifesto against "pure consciousness" as it was the statement of a program for investigating precisely the relationships between consciousness and that which it is directed towards. Beginning with a perception model, he attempted first to sketch out the various kinds of relationships between perceptual consciousness and the perceived "object." The same "object" was then revealed as the intersecting point of multiple noetic acts directed towards it, the "filling out" of discrete intentional acts of seeing, remembering, and so on is provided by these multiple noetic perspectives, and the "object" is identified as the "noema" of perception, remembrance, phantasy, etc. The noematic core of the convergence of all these intentions could be called "essence."

Too much has been made, perhaps, of Husserl's use of the term *Abschattungen* to express these various perspectives, sketches, adumbrations. All of these terms have been tried and found wanting; the present tendency is to leave the term untranslated. In any case, this is not *the* method of phenomenology, nor is the Husserlain *Wesenshau* (the famous "intuition of essences"). These procedures are intermediary, or even preliminary, stages in phenomenological investigation, serving to clarify the field of research and provide guidelines along which to proceed toward the self-givenness of objects of perception and thus get "back to the things themselves." These procedures are only steps in the "filling-out" of our "intentions" with regard to that which is to be known.

In *this* (preliminary) sense of getting "back to the things themselves," Husserl simply sketched out some of the more obvious perceptual relationships between noema and noesis, and suggested that the same procedure could be used to identify the relationships between loving and loved, affirming and negating and that which is affirmed and negated, between willing and willed, and so on, so as someday (hopefully) to achieve a complete inventory of conscious activity. *Acts* of consciousness, of course, were in view; but an act without content is at best unstructured and unnamable and at worst simply empty or nonexistent; consequently,

[8] Rancurello (116-117) succinctly presents the relationship between Brentano and phenomenology: "In the main, the Austrian school got its phenomenological inspiration from Brentano; the Würzburg school from Husserl who had taken it from Brentano; Gestalt psychology from the Würzburg school and from Stumpf, who also had taken it from Brentano; and Jaensch and Spearman from Husserl."

acts had to be considered as structured intentionally in relationship with the various objects of intention. Husserl continued his own researches in the direction of transcendental intersubjectivity and his final radical critique of Western science in its turn of the century crisis. Others took up the task of explicitating the concrete *a priori* of the life-world and its relation to conscious subjectivity: Scheler in his study of loving and hating, Ricoeur in his work on the act(s) of willing, and so on. Still others continue the basic reflections of Husserl in the explicitation of the life-world and its intentional *a priori*, and of the intentional intersubjectivity which makes of the "universe" a world of human persons in a human landscape. This is not the place to attempt an account of their efforts; Spiegelberg can best guide the reader here, and the increasing library of phenomenological literature in English—both translations of original texts and new studies—will provide a solid basis for phenomenological reflection.

The work of Husserl could not go unmentioned in these paragraphs; but consideration of it is necessarily summary. The writer considers Husserl's work primarily epistemological and ontological, forming the basis for all the thinkers and researchers who are attached to the tradition of phenomenology; but, as such, it is not another system among philosophical systems. It is, rather, a method and an attitude which provides a basic critique of both philosophy and science, and is the motivation of all those who attempt to re-think radically the place of scientific and philosophical models in search for the meaning of man. For this reason, we shall consider the phenomenology of Husserl, in its radicalness, and the use of it by those who carry forward his enterprise, as a transcendental phenomenology. We shall not pursue this, but rather we shall consider a branch of phenomenology which is equally important in the formation of what Maslow has called contemporary third force psychologies: that which takes its origin from Brentano, as did that of Husserl, but passes through the *empirical* phenomenology of Stumpf. (A *de facto* recognition of the possibility of this kind of effort can be found in the title of Straus' series, "Phenomenology: Pure and Applied.")

Contemporary third force psychology seems to have little resemblance to the work of these precursors; and it has not succeeded in really establishing a viable paradigm to compete with mainline experimental psychology.[9] However, it is gaining adherents rapidly on the basis of its

[9] Amedeo Giorgi, in his *Psychology as a Human Science: A Phenomenologically Based Approach*, New York, 1970, has sketched out the requirement of such a paradigm. The writer's review of this book will be published in the international journal, *The Human Context*, London, 1971.

ideology. It can be taken for granted that third force psychology is now embodied in the Association for Humanistic Psychology, and that the Association includes academic as well as clinical psychologists, encounter groupies, and, in general, everyone who ascribes to Floyd Matson's description of the movement as it surfaced in the early 'sixties:

> An affirmative approach which dared to talk of the human resources of possibility and promise of joy and love and simple growth—and even (reverting to a forgotten myth) of the active self, the self as subject.[10]

The third force also adopted a critical stance against what it considered man's inhumanity to man in the experimental laboratory and on the analyst's couch: "the psychogenetic determinism of orthodox Freudians which made man the creature of instinct and the environmental determinism of the behaviorists which made man a creature of conditioning."[11]

Thus, within the third force, the key concepts guiding all activities in favor of human personhood have been those of possibility, potentiality, growth, selfhood. However, there is a discernible ambivalence regarding the relative importance to be accorded to the experiencing of these as realities rather than thinking of them as concepts. The source of this ambivalence may be twofold: First, there is perhaps the conviction that reliable scientific verification of the "experiencing" in question cannot be justified in the current sense of the term "scientific," or is even impossible (given the fact that there is no recognition of the possibility of verifying *over time* a being which is constituted *by time*.) Second, there may be the psychologist's traditional distrust of philosophy. After all, the whole profession of mainline psychology, following Wundt, has been historically a movement of liberation from philosophical method in the direction of the methods of the physical sciences. If the third force psychologist has renounced the paradigms of the natural sciences, can he now ask for philosophical justification and still claim to be a psychologist in any sense of the term? He is the "third man" in more than one sense; his requests for theoretical justification (though no theory is acceptable) and for philosophical respectability (though philosophy is traditionally anathema) can only be half-hearted. And where is he to go? Back to the APA, as some are urging? Or out into the void, as some are actually doing? Yet, unless something happens soon, the odd man will be out. Let us be frank: the third force is at a perilous crossroads, and

[10] Floyd W. Matson, "Whatever Became of the Third Force?" (Presidential address at the AAHP Seventh Annual Meeting, August 29, 1969), *Newsletter: American Association for Humanistic Psychology,* 1969 (6).

[11] *Ibid.*

serious psychologists/philosophers are engaged in a duel with crass charlatans for the discovery of the true character of the third force.

A resolution of the problem might well center upon a resolution of the basic ambivalence in the organized third force.

A resolution of the problem of philosophy in third force psychology has already been begun by Amedeo Giorgi in his book, *Psychology as a Human Science: A Phenomenologically Based Approach,* in the International Association for Phenomenological Psychology which he founded, and in the journal which he edits for it. Giorgi has neatly shown that the kind of philosophizing against which Wundt rebelled is not at all likely to return to threaten the contemporary psychologist. Wundt was turning away from the method of deduction and speculation—that is, the method of classical systematic philosophy—and turning toward the method of induction and the empirical establishment of facts. Giorgi has also shown that a new kind of philosophical enterprise could be of enormous service both to philosophy and to psychology; and he suggests that, in fact, a "reflective method" is already at work in any psychological enterprise.

Philosophically speaking, there must be a return to, or a rediscovery of, the foundations of third force psychology; its problems would never have emerged as problems without the efforts of the precursors, but these have been largely forgotten. If today the third force is contributing

> to a rehabilitation of the self in its subjectivity and hence in its dignity, to a replacement of the mechanomorphic model of the robot with the anthropomorphic model of the person, and has turned attention from determinism to self-determination, from causality to purpose, from behavior to experience, from manipulation to actualization,[12]

there is a long underground philosophical tradition supporting the effort, and the problem was set almost a century ago.

The fifty-year vendetta between the "Leipzig School" (the friends of Wundt) and the "Würzburg School" (the friends of Brentano) is well known; the Leipzig School finally won the battle, but the Würzburgers did not give up. Brentano's *Habilitation* in Würzburg and its influence upon Stumpf has been mentioned; Külpe, with his bipartite psychology, continued the battle in Würzburg. The late Karl Bühler was Külpe's student, and so was Max Wertheimer—and Kurt Lewin was a student of *his*. The history of the Würzburg School cannot be traced here, but it is possible to map the milestones which lead from Brentano, the grandfather of phenomenology, to our own day's third force in psychology

[12] *Ibid.*

which is his intellectual grandchild, even though the intermediate parent was Stumpf rather than Husserl. If it has been forgotten that both Husserl and Stumpf were the prominent students of Brentano, it is even less well remembered that Wertheimer, Koffka, and Köhler, founders of Gestalt psychology, were students of Stumpf at one time or another. It is probably unknown that Kurt Lewin, as a young man, was also a student of Stumpf. It is from Lewin, perhaps, that we have the most enlightening statement linking the third force in psychology with the tradition of Brentano, Husserl, and Stumpf. He wrote in 1937 that Stumpf, in his day, had been particularly representative of the famous transition of psychology from a philosophical status to that of an empirical science. Lewin explained that Stumpf's training with Brentano had led him to the "quasi-empirical philosophy" which Spiegelberg has recognized—the same one which led Husserl to phenomenology. Lewin claimed that according to all three of these precursors, the true philosophical method was identical with that of the natural sciences. Here the cleavage in the tradition of psychology is most apparent: Wundt's people were renouncing deduction and speculation and turning toward induction; Brentano's people refused the either/or and opted for the complementarity, even the *compenetration* of inductive and deductive methods, attempting a totality of ways of knowing the totally human phenomenon.

The Gestalt psychologists, then, were the disciples of Carl Stumpf, who had his ideal of the relationship of science and philosophy from Brentano. Before 1920, Lewin had been a student of Stumpf and of Wertheimer, and a classmate of Köhler. He was later to say that the fundamental ideas of Gestalt psychology were the basis of all his investigations in the area of will, affectivity and personality. Lewin's Field Theory was a direct outcome of the theory of Form; and as far as the author knows, he was the founder of the first non-therapeutic group experience setting.

Stumpf's disciples had other effects upon American psychology too. It is very instructive to notice, for example, that Gestalt psychology was influential in the work of Donald Snygg and Arthur Combs. Snygg wrote a paper entitled, "The Need for a Phenomenological System of Psychology," in 1941. Snygg was in a good position to observe the shift in emphasis from behaviorism to phenomenology in American psychology. He had begun his career as a behaviorist, but had developed a phenomenological point of view, he said, under the influence of Lewin and Köhler. Snygg found a kindred spirit in one of Carl Rogers' students, Arthur Combs; and together they produced *Individual Behavior: A New Frame of Reference for Psychology* (1949, 1959). Gestalt psychology, and the theoretical systematization of it by Snygg and Combs in the "behavioral sciences" are considered to be the background for the work of Carl Rogers, one of the recognized leaders of third force psychology today.

Another link in the great chain of the historical being of the third force was Abraham Maslow—and oddly enough his teachers were Wertheimer and Koffka, after their emigration to the New School for Social Research in New York in the early 'thirties. Maslow's important work began, we know, as early as 1935. In his words, it was "an effort to synthesize the holistic, the dynamic, and the cultural emphases" beginning to emerge in American psychology at the time.[13] The bridge between the holistic and the dynamic, he asserted, had been built for him by Kurt Goldstein.[14]

The only justification for brushing this large canvas of the history of the third force (besides the lack of space to fill up the gaps) is that it provides a background for the thesis of the phenomenological origins of this movement and a suggestion of the connectedness of the basic theories and options which influenced it. First, there is Lewin's *Field Theory* (incorporating the idea of the perceptual field and the self within it). Second, there is *Holism* (linking the perceptual field with the biological realm of real bodily existence). Third, there is *Gestalt* (developed philosophically by Merleau-Ponty in the direction of the form of the embodied self against the background of the world, and developed clinically by Carl Rogers in the direction of the perceived form of the self against the background of society and culture). Finally, there is Maslow's genial transformation of the whole enterprise of clinical psychology (regarding individuals or groups) from a medical model to a *growth model*, following Goldstein. In general, this backdrop, with all its coloration, is dependent in one way or another upon the phenomenological movement, beginning with Brentano and dividing into complementary streams headed each by Husserl and by Stumpf. A remembrance of this tradition, particularly Brentano's conviction of the complementarity of what we would call the "inside view" and the "outside view," might solve the philosophical problem of the third force today.

The solution of the philosophical problem might also provide a solution to the scientific problem; for here again it has been the phenomenological philosophers who have helped explicitate what sort of being is *human* being when released from his status as a simple object in the world. The

[13] Abraham Maslow, *Motivation and Personality*, New York, 1954, 5.
[14] Kurt Goldstein, *The Organism*, New York, 1937.

connection between Brentano's psychology and Husserl's phenomenology is becoming clearer year by year. Husserl's use of the idea of consciousness as "consciousness of" made it possible for him, early in his career, to make the breakthrough distinction between "objective" time and the constitution of temporality in human consciousness.[15] Heidegger went even farther and attempted to present "an interpretation of human existence in terms of temporality..." [16] In other words, the oversimplified and externalized concept of objective time is not one in which we live as human subjects. The time in which we exist is an extremely complex criss-crossing of experienced relationships. Second, it is clear that our temporality is not a being, but a process which "temporizes." The constituent moments of this process are not three parts; they are three different directions in which this process comes to pass.[17] Third—and most important here—we, as human beings, are constituted not as objects-in-the-world suitable for natural scientific verification; we are subjects essentially constituted by time.

Where are these reflections leading? To the statement of a theory which might help solve the problem of evolving a defensible concept of the human person in third force psychology. The elements are there: It was noted that we, as persons, are constituted by time (it is also the case that our temporality is not discontinuous, but space does not permit substantiation of this important point); then, there are the chief theoretical contributions to third force psychology, Field Theory, Gestalt, Holism, Growth. And if the reader will grant the argument that third force psychology was set in motion by the phenomenological concerns of the precursors, then perception must be added as a basic integrating concept. This adds up to *a self theory based on perceptual principles, especially those pertaining to the perceptual field of the self, in which the self appears to the self as a gestalt. Since this self is a human organism, it must be understood in a holistic sense.*

This theory is already in existence, and has been worked out extensively by Carl Rogers.[18] Rogers' theory incorporates all the basic options of (phenomenological) third force psychology, subsuming them all under the single integrating nuclear concept of *the self.*

[15] Edmund Husserl, *The Phenomenology of Internal Time Consciousness,* ed. Martin Heidegger, trsl. James S. Churchill, Bloomington, Indiana, 1964.

[16] This is part of the subtitle of Martin Heidegger's *Sein und Zeit (Being and Time),* 1927.

[17] This is William Richardson's explicitation of Heidegger's thesis of human being's temporality, *Husserl: Through Phenomenology to Thought,* The Hague, 1968.

[18] Carl R. Rogers, "A Theory of Therapy, Personality, and Interpersonal Relationships as Developed in the Client-Centered Framework," *Psychology: A Study of A Science,* Vol. III, *Formulations of the Person and the Social Context,* ed. Sigmund Koch, New York, 1959.

Contributing to this single focus of Rogerian theory are, first, the organism, the total human individual; second, the perceptual field, which is the totality of experience; third, the self, a differentiated segment of the perceptual field. The self is constituted by a constellation of conscious perceptions and values of the "I" or "me," a gestalt more or less available to awareness and subject to the same kinds of transformations as all gestalten. This organismic totality is said to have the following properties: It relates as an organized whole to the *world as perceived* (i.e. the phenomenal field, not necessarily "the real world," whatever that may be). It has one basic complex motive, which is to maintain, enhance, and actualize itself. It may symbolize experiences so that they become conscious, deny them so that they become or remain unconscious (or "subceived") or distort them so that they come into conflict with the real existential experiencing of the organism.

Rogers' total theory is a far-reaching and coherent personality theory, the explicitation of which is impossible in this limited space. How does it solve the philosophico-scientific problem set above? It is plain that the answer must attempt to reconcile the old dichotomy which has plagued psychology from its beginnings and continues to cause discomfort: which comes first, experimentation or theory? Rogers' theory of the self, the core of his total theory is the result of extensive and highly sophisticated operationally defined scientific procedures [19]—in other words, the kind of scientific method which ought not to be applicable to human selfhood understood as a process of becoming. On the other hand, the theory falls within the third force tradition which is here insisted upon as phenomenologically inspired; and it is expressed precisely in terms of possibility, potentiality, growth, personhood, selfhood.

In answering the problem above, one must note that experimental data produced in support of hypotheses in Rogers' context are collected in an attempt to verify a hypothesis based on some experience *outside* the experimental situation. Data, then, help support the conviction that my intuition about what has occurred is more or less correct; but data do not reveal that experience to me, nor do they demonstrate all the dimensions of that experience. Natural scientific method yields results which are less rich but more sure, and thus seem "closer to reality;" philosophical reflection may yield conclusions which transcend even

[19] Rogers was heavily indebted to William Stephenson, *The Study of Behavior: Q-Technique and its Methodology,* Chicago, 1953. The most extensive report of research using this technique is Carl R. Rogers and Rosalind F. Dymond, *Psychotherapy and Personality Change,* Chicago, 1954.

statistical certitude, and may be equally "real," though not mathematically verifiable. These two methods—scientific and reflective—are separate moments in the search for human meaning and each has its particular excellence; but it does not help to confuse them, nor does it help to forget that what one is attempting to describe, verify, and reflect upon is a primordial phenomenon—neither scientific nor philosophical, but the source of possibility that there *be* science and philosophy. The reader's attention is called earnestly to this point; it may well express the core-message of this essay. From the time of Plato we have been warned of the illusory quality of our everyday experiencing; philosophy attempted to de-mystify our supposed originary intuiting of reality, and it failed. Science has further attempted to do so, and has so far failed—at least for third force psychology. What science and philosophy have positively gained in the explicitation of our primordial experiencing cannot be negated—unless scientific and philosophical elaboration become *substitutes* for that originary intuition of reality. What is being urged here, in support of the constitution of a holistic paradigm of human being, is the possible complementarity of scientific verification and philosophical reflection, together with a disciplined subjectivity which could make possible the conscious elucidation of primordial experiencing, and a transcendental focus which could make evident the *a priori* conditions of possibility of that elucidation.

Third force psychology, although it has forgotten its late nineteenth century origins, is carrying forward a tradition begun by Franz Brentano, who refused the either/or stance of Wundt's psychology. Since third force people are plagued by this forgotenness, it is hoped that a recognition of the paradigm-character of Rogerian theory can help them become, if they wish, both scientifically respectable and philosophically relevant. Rogers has, in fact, demonstrated that one can verify over time a being who is constituted by time. He has bridged the gap between scientific psychology and the phenomenological bases of third force options. The path is now open for conscious and systematic reconciliation of the behavioral sciences and philosophical reflection; but the only conscious strategy which will work for the third force is the one which the philosophical precursor of that movement invented, that is, Edmund Husserl's phenomenological psychology. Before 1913, Husserl had achieved a viable distinction between classical experimental psychology and his phenomenological psychology. Husserl's eidetic psychology, as Joseph Kockelmans well said, "was to provide a foundation upon which a true experimental psychology could be based; it would also bridge the gap between experimental psychology and philosophy." The writer would like to suggest further that, beyond bridge-building, these two strategies are "nested" one within the other and constitute together the complementarity and compenetration of the "inside view" and the "outside view" which Brentano saw as the only way of explaining the total phenomenon of being human. This gives us, in the writer's opinion, the sole access we have to the *foundations of both science and philosophy* in the everyday experiencing of being human.

But a study of foundations does not stop with foundation. It gives meaning to the enterprises which are based on them. In this way, third force psychologies, insofar as they are phenomenological, have been able to avoid the "nothing but" reductionism of so much of our contemporary endeavors, and (with the tolerance which Boring grants them) have been able to incorporate other disciplines as well. Chief among these is a biological tradition. In America, the early days of most of our sciences were influenced in various ways by Darwinism (we shall see its effects in sociology, for example).

On the milestones in the contemporary tradition of third force psychologies, can be read the names of James and Dewey, with intermediate states marked evolutionism, transactionalism, and functionalism. This position owes quite as much to Darwin as to any psychologist, and has the merit of a remarkable consistency. One of the best inclusive efforts at harmonizing psychology and biology is found outlined above in the theory of Carl Rogers. Rogers finds, on the basis of the question, "If the individual is what he is revealed to be in therapy, what kind of theory will account for him?" that the individual human being is provided with two interacting growth systems, an "inherent motivational system" which all living beings possess (an "actualizing tendency") and a "regulatory or feedback system" (an "organismic valuing process"). The organismic penchant here is striking; Rogers might, for example, have used a cybernetic model with equal clarity, taking the mechanistic viewpoint. Ashby has already shown that the robot-machine, precisely because provided with a feedback system, is even theoretically capable of extensive adaptation. On this basis, one could have built a mechanistic model even of human freedom; but that is not the penchant of "humanistic" psychologists. Put in simple schematic terms, the key idea is that the individual is capable of inherently regulating his growth and enhancement by the control of his experience. The person functions as a system of regulated goals, the organism being oriented primordially toward its own maintenance and enhancement, and provided with a regulatory system whereby it evaluates the results of its activity in relation to the ends which it pursues. The evaluation serves as an orientation for subsequent activities.

In classical terms, this might be considered an instance of entelechy. According to Aristotle, one *logos* of a being is τὸ τί ἦν εἶναι, that is, what the being in question "was to be." (*Meta.* A, 3, 983 a 28.) Insofar as this *logos* names a principle of being, it refers to an intrinsic principle of *becoming*. The being becomes what it "was to be" according to the principle by which it already is. Applying this to a living being, he finds that the soul (form, or rather, nature) is "a primary entelechy of a natural body having life potentially, that is, of an organized body." (*de An.* II, 1, 412 a 27.) Entelechy, then, is a concrete instance of what the being in question was to be, that is, a nature or principle by which the organized body becomes, as we say "what it is."

For the "scientific" biologist, "entelechy," as Dobzhansky pointed out, was nothing but a horrid word, empty, explaining nothing; and it fell under the condemnation of the sworn pact of 1845. Hans Driesch, the eminent embryologist, who died in 1941, was the last great man who admitted being a vitalist. Today, remarks Dobzhansky, "every student taking an introductory biology course in expected to learn that mechanism is right and vitalism is wrong." "However," he goes on, "not every student, and not even every professor, has a clear understanding of why this is so." [20] We must leave to the reader the occasion to try out the details of the argument; Dobzhansky and von Bertalanffy are the best guides here. The former's *The Biology of Ultimate Concern*, in the "Perspectives in Humanism" series planned and edited so capably by Ruth Nanda Anshen falls within the thesis of this book.[21] Von Bertalanffy's *Robots, Men and Minds* [22] is an effort, on a grand and all-inclusive scale, to "humanize science" with a new image of man and a new natural philosophy, surpassing by integrating the organismic concept in psychology and biology with a new natural philosophy which itself surpasses

[20] Theodosius Dobzhansky, *The Biology of Ultimate Concern*, New York, 1967, 19.

[21] Miss Anshen writes in commentary: "However far-flung its hypotheses or comprehensive its theories, science has no objects for its application save such as can be known through humanistic interpretation ... What this Series hopes to demonstrate is that humanism by its nature is intent upon forcing the mind to make, since it is unable not to make, judgments of value." (*Ibid.*, xiii). The writer would add that one of the basic positions of the present volume is that such judgments are also made in the very *creation* of the scientific enterprise, and not only in its application.

[22] Ludwig von Bertalanffy, *Robots, Men and Minds: Psychology in the Modern World*, New York, 1967.

the limitations of cybernetics and even general systems theories such as his own in the direction of total openness and creativity.

A choice has evidently to be made. In a model based on evolutionary biology, we can go in either direction: toward Weston La Barre's concept of "the human animal," in which "A billion other worlds may turn, without end and without meaning, in the cold cosmic night. But on this one earth, at least now live animals able to become even a little like gods, having knowledge of good and evil," [23] or toward Desmond Morris' *The Naked Ape*,[24] which was written as a direct insult to these pretensions, on a purely ecological model and in the tradition of ethology.

What must be done is to calm the scientist's fears about not being properly scientific, by recalling that, phenomenologically speaking, *all* views of man may be true, but *only as a totality and not as a sum,* since man himself is a totality and not a sum. It is the option of the "organismic" penchant in psychology and biology that the totality of man is a given. The problem, of course, is the question of a *telos*. Teleology creates for the scientist a problem which cannot be solved in either mechanistic or operational terms; it implies that, somewhere, there is an *ought* in human existence, but scientific technology cannot deduce it from the *is* which is its concern. Landgrebe asserts that Driesch's espousal of "entelechy" simply "expressed the embarrassment which is caused by the abortive attempt to explain life mechanistically."[25] But even in the rejection of the theoretical aspect of Driesch's discoveries, he notes, there are irrefutable experimentally secured findings, among which are "the facts of *self-organization* and *self-differentiation*—facts which exclude any pre-determination and make it mandatory to look for the factors by which this self-organization is directed." And Landgrebe agrees with practically all the other authors cited here that it is also irrefutable that "vital impulses of this direction are to be sought in the realms of the micro-physical." [26] It is interesting to note, in addition, that even Hans Jonas, in his book *The Phenomenon of Life,* while taking issue with the evolutionary optimism of Teilhard de Chardin which Dobzhansky seconds, must find place for a discussion of the recent findings of micro-biology regarding the DNA molecule. The set of chromosomal DNA molecules which represents the genetic code of each living cell and controls the metabolism of the cell is, itself, not subject to that metabolic exchange,

[23] Weston La Barre, *The Human Animal,* Chicago, 1954 (6th ed. 1965) 334.

[24] Desmond Morris, *The Naked Ape,* New York, 1969.

[25] Ludwig Landgrebe, *Major Problems in Contemporary European Philosophy: From Dilthey to Heidegger,* trsl. Kurt. F. Reinhardt, New York, 1966, 89.

[26] *Ibid.*

and this prevents an easy negation of "an internal principle of identity" in physical particles.[27] However, its significance is not under discussion here; [28] Dobzhansky, von Bertalanffy, and Jonas draw conclusions about it pertinent to their search for an integrating vision of man.

We shall keep our attention on the more accessible statement of Rogers, as a model of the holistic-biological-psychological character of third force psychologies. Everyone insists upon the actualizing tendency of the living organism as an *innate* principle of growth, although the character of this innateness must be searched for our present purposes. It is, in general, "the innate tendency of the organism to develop all its capacities in ways which serve to maintain or enhance the organism." [29] Not only is the tendency innate; it is also *directional* in character. It is "evident in the life of the organism from conception to maturity, at whatever level of organic complexity." [30] Here ontogeny is viewed as recapitulating phylogeny; the directional trend noticeable in the evolution of the individual is an individual sample of the observable trend at all levels in the process of evolution, and generalized differences observable in the evolutionary scale indicate the deployment of this fundamental growth principle. On this point, a whole cluster of varying points of view come to focus. The basic notion of the directional tendency in organismic actualization in this context, is borrowed, of course, from Kurt Goldstein, who was influenced by Gestalt theory.[31] A nuanced view was advanced by Andras Angyal.[32] Recently rehabilitated because of the renewal of interest in philosophical biology, Angyal contended that the organism's strivings and developments are "not strictly teleological but merely directional." Angyal's work may yet provide the bridge between the beleaguered scientist and the anxious

[27] Hans Jonas, *The Phenomenon of Life: Toward a Philosophical Biology*, New York, 1966, 82, 97-98.

[28] "Biological Philosophy" is emerging as an important phase of contemporary reflection upon the nature of human being-in-the-world. Its considerations are more complex and sophisticated than can be managed in an introductory essay. A study of this model must be deferred for treatment elsewhere, as must the equally sophisticated "General Systems" approach. Even the relatively simple models under scrutiny here have many interdisciplinary influences and borrowings which have gone unmentioned in an effort to keep each model relatively clear and distinct. It is hoped that the reader has identified some of these for himself.

[29] Rogers, "A Theory of Therapy," 196.

[30] Rogers, *Client-Centered Therapy*, 488.

[31] Kurt Goldstein, "Notes on the Development of My Concepts," *The Journal of Individual Psychology*, 1959 (15), 5-14.

[32] Andras Angyal, *Foundations For A Science of Personality*, Cambridge, Mass., 2nd ed., 1958. Angyal, with Goldstein, is one of the chief proponents of an "organismic" theory of personality, which, according to Hall and Lindzey, is "the extension of Gestalt principles to the organism as a whole". The chief philosophical proponent of organismic theory was Jan Smuts, in his book *Holism and Evolution* (1926). Smuts coined the term "holism," and Angyal, following Smuts' ideas, invented the concept of the "biosphere."

humanist. He did not deny the teleology of organismic activity tending toward fixed goals, but claimed that this sort of teleology is only secondary. "*It is not the goal which defines the direction,*" he said, "*but on the contrary, the intrinsic pattern of a direction which defines what object can become a goal.*"[33] In view of the problems which scientism has created for ethics, moral theology, and theodicy, it seems that discussions of "natural law" might profitably begin by reflection upon this statement.

The innate directional growth here spoken of is usually understood on a vast scale. It surpasses what Maslow terms "deficiency needs" for air, food, water, and the like; and for Rogers it involves more generalized functions: "development toward the differentiation of organs and of functions, expansion in terms of growth, expansion of effectiveness through the use of tools, expansion and enhancement through reproduction. It is development toward autonomy and away from heteronomy, or control by external forces."[34] This statement of the vastness of the directional power of growth must also be extended to the growth of the human self which can be seen to be continuous with the evolutionary model given here. In that context (which, unfortunately, cannot be developed) self-enhancement surpasses, in the same way, the mere maintenance of selfhood. Rogers, I think, sums all this up nicely: "...the organism is an active initiator and exhibits a directional tendency." He quotes the appealing judgment of R.W. White, who said, "Even when its primary needs are satisfied and its homeostatic chores are done, an organism is alive, active, and up to something."[35]

It is possible, of course, to consider individualized aspects of this basic directional tendency, and one can admit motivational concepts such as "need-reduction," "tension-reduction," "drive-reduction," and others dear to rat fanciers; but growth motivations which go beyond these terms and are inexplicable by them (like "the seeking of pleasurable tensions, the tendency to be creative, the tendency to learn painfully to walk when crawling would meet the same needs more comfortably") must be insisted upon even more strongly.[36] None of the tendencies ordinarily called

[33] *Ibid.*, 53-55.

[34] Rogers, "A theory of Therapy," 196.

[35] Rogers, "The Actualizing Tendency in Relation to Motives and to Consciousness," *The Nebraska Symposium on Motivation*, 1963 (ed. Marshall R. Jones), Lincoln, 1963, 5; R.W. White, "Motivation Reconsidered: The Concept of Competence," *The Psychological Review*, 1959 (66), 315.

[36] Rogers cites Goldstein on this point: "The tendency to discharge any tension whatsoever is a characteristic expression of a defective organism, of disease." (Kurt Goldstein, *Human Nature in the Light of Psychopathology*, Cambridge, Mass. 1947, 141.) (See also Gordon Allport, *Personality and Social Encounter*, Boston 1960, 107.)

motives, in any case, can be considered autonomous. The actualizing tendency of the organism is not a combination of instincts, reflexes, or needs, but rather the polyvalent but unique drive of which these are simply manifestations.

The actualizing tendency, then, is the only motive which Rogers will postulate in his theory. Thus, he says, "it is the organism as a whole and only the organism as a whole, which exhibits this tendency. There are no homunculi, no other sources of energy or action in the system." [37] Here Rogers is in agreement with Maslow. On the point in question, Maslow thought that the "instinct theory" may have to be reexamined and perhaps rehabilitated on the basis of current theory and research regarding the actualizing tendency.[38] A dissenting opinion was voiced by Gordon Allport, who argued that the human personality is more a function of development than of instinct. His oft-repeated view is based on his own theory of "functional autonomy." His famous brother, Floyd Allport, the social psychologist, has done the best job of unifying these diverse tendencies in his own theory of "dynamic structure." [39]

On this point, then there is agreement that the actualizing tendency of any organism, and of man as a total organismic unity, is innate and directional, transcends mere homeostasis in the direction of enhancement, is universal and all-pervasive in the organism, and is the unique but polymorphous motivational principle. What was originally found to be mere "lawfulness" has turned out to be a "purposeful lawfulness" as well.

As this group refuses the atomization of the human personality, so it refuses the atomization of human behavior. Rogers, whom we take here as spokesman, stated (in explaining the behavior of the human infant) that he "interacts with his reality in terms of his basic actualizing tendency. Thus his behavior is the goal-directed attempt of the organism to satisfy the experienced needs for actualization in the reality as perceived. In this interaction he behaves as an organized whole, as a gestalt." [40] When the stage in human development is reached where the infant begins to form a concept of selfhood, with its concomitant sense of personal value and learned cultural values, the meaning of the term "organism" is extended beyond the limits of biological individuality and defines a specifically *human* individual. While Piaget may not quite "belong" in this genealogy,

[37] Rogers, "A Theory of Therapy," 196.

[38] Abraham Maslow, *Motivation and Personality*, New York, 1954, 123-145.

[39] Floyd H. Allport, *Theories of Perception and the Concept of Structure: A Review and Critical Analysis with an Introduction to a Dynamic-Structural Theory of Behavior*, New York, 1955.

[40] Rogers, "A Theory of Therapy," 222.

his lifework has been devoted to the study of the successive waves of development in children, and his work has been comprehensively and beautifully systematized in Guilford's major contribution. *The Nature of Human Intelligence*. Guilford's book shows in a remarkable way how a basic view of man which outruns both traditional associationism (inherited from the 18th century) and behavioristic stimulus-response theory (the first commandment of the 20th century) can incorporate and transcend all mechanistic views of man and his world.

There is a second organismic process in the theory of Carl Rogers which ought to be mentioned, if only briefly, and that is an organismic *valuing* process. Conceived on the model of a feedback system by which the individual regulates his behavior and controls his choices, the organismic valuing process is the goal-directed process by which the individual evaluates his experiencing. The use of the concept of an organismic valuing process as a regulatory system in the context of a theory of personality is of recent origin, and there is little material with which to work. The postulate of such a valuing process is an inference from the context of outcomes and processes of psychotherapy, in which the problem of valuing first became meaningful. I think the valuing process is much better understood in that context as well (for example, in Rogers' notion of the shift in the "locus of evaluation" as an outcome of therapy); but space does not permit that here, and a few remarks must suffice.

Evidently the concept of an "organismic valuing process" owes much to Goldstein and Cannon, and even before them, to Claude Bernard.[41] We are indebted to Cannon for coining the term "homeostasis" to indicate the maintenance of a steady state in the organism. He studied this phenomenon especially in terms of the function of endocrine secretions. Even more significant were the studies of Cannon and others regarding *total* organismic responses as governing the maintenance of steady states.[42]

Cannon himself thought he had disengaged *general* principles of stabilization, and employed analogies based on industrial, domestic, and social organization.[43] Goldstein, while being careful to indicate his dissatisfaction with "physical gestalten" in the isomorphic sense of the later Gestalt theory (Goldstein had discovered the still inexplicable phenomenon that

[41] Claude Bernard, *Leçons sur les propriétés physiologiques et les altérations pathologiques des liquides de l'organisme*, Paris, 1859. Walter B. Cannon, *The Wisdom of the Body*, 2nd ed., New York, 1963. Kurt Goldstein, *The Organism*, 2nd ed., Boston, 1963.

[42] C.P. Richter, "The Internal Environment and Behavior: V., Internal Secretions," *The American Journal of Psychiatry*, 1941 (97), 878-893. C.M. Davis, "Self-Selection of Diets by Newly Weaned Infants," *The American Journal of Diseases of Childhood*, 1928 (36), 650-656.

[43] Cannon, 305-324.

the classical cortical functions could be transferred to other cortical loci), nevertheless continued the application of the notion of homeostasis to the *total* behavior of organisms, and described "good gestalt" as "a definite form of coming to terms of the organism with the world." [44] Straus' interest in the phenomenon of the "phantom limb" goes in the same direction.

The effort of this chapter has been to indicate the *use* of phenomenological principles derived from the phenomenological method of Husserl, employed in the empirico-phenomenological tradition of Stumpf, and best exemplified in the contemporary theory of Carl Rogers. This point has been insisted upon since that is where the present thrust of phenomenological method in psychology is to be found. It is quite different from that of Binswanger, for example, whose description of the *Lebenswelt* of his psychotic patients does not directly serve in the process of therapy. It is compatible with the philosophical effort of Lawrence and O'Connor, precisely because their preoccupation with meaning *and* value unites philosophically their interest in phenomenology (which is primarily an epistemological method) and in existentialism (which is primarily an ethics based upon phenomenological understanding of human existence). It seconds the "phenomenological psychology" of Minkowski, Straus, and Gurwitsch, and their disciples in the tradition of Husserl which Joseph Kockelmans has recently put so clearly and solidly in focus. It shows less respect, in general, for current American efforts to "phenomenologize" empirically while denying the transcendental reduction which would enable them to do so.[45] This kind of effort at explicitating the *Lebenswelt* lacks the solid phenomenological underpinning of the others, giving the impression that *to do phenomenology is to describe*. Such a phenomenology is indistinguishable from the pure empiricism which the phenomenological enterprise has struggled so fiercely against. In the context of a holistic model of human-being-in-the-world, this book has attempted to gather under the canopy of a common tradition all those who have contributed, from many perspectives, to an empiricism which accepts as truly human that *pre-knowledge* we have of ourselves and of our human being which Husserl, under the inspiration of Brentano, attempted to justify, and which Stumpf in the wake of Husserl, carried through. Furthermore, this book attempts to respect the "hermeneutic spiral"

[44] Goldstein, *The Organism*, 369-397, esp. 371.

[45] See my comparative review of Johannes Linschoten's *On The Way Toward a Phenomenological Psychology: The Psychology of William James* and John Wild's *The Radical Empiricism of William James*, in *The Human Context* (London), Winter 1970, 147-153.

inherent in the recognizably double enterprise of transcendental and empirical phenomenology—that in order to describe one must first understand, and that description of what is understood (even when empirical means are used) must follow the structures of understanding disengaged by phenomenological reflection. Brentano's original insight of the possible coordination of phenomenological and explanatory procedures has been, the author believes, substantiated and (at least in an inchoate fashion) exemplified.

SELECTED ADDITIONAL READINGS

FOR CHAPTER II

Allport, Gordon W., *Becoming: Basic Considerations For a Psychology of Personality,* New Haven, 1955.

Arnold, Magda B., *Emotion and Personality,* Vol. I: *Psychological Aspects;* Vol. II: *Neurological and Physiological Aspects,* London, 1960.

Bannan, John F., *The Philosophy of Merleau-Ponty,* New York, 1967.

Barron, Frank, *Creativity and Psychological Health* (in paper: *Creativity and Personal Freedom*), Princeton, N.J., 1963.

Binswanger, Ludwig, *Being-in-the-World: Selected Papers of Ludwig Binswanger,* trsl. Jacob Needleman, New York, 1963.

Blake, Robert R., and Glenn V. Ramsey, *Perception, An Approach to Personality,* New York, 1951.

Brentano, Franz, *Psychology from an Empirical Standpoint,* trsl. Antos C. Rancurello, London, 1968.

Bugental, James F.T., ed., *Challenges of Humanistic Psychology,* New York, 1967.

Combs, Arthur W., and Donald Snygg, *Individual Behavior: A Perceptual Approach to Behavior,* 2nd ed., New York, 1959.

Gendlin, Eugene T., *Experiencing and The Creation of Meaning: A Philosophical and Psychological Approach to the Subjective,* Glencoe, Ill., 1962.

Giorgi, Amedeo, "Existential Phenomenology and the Psychology of The Human Person", *Review of Existential Psychology and Psychiatry,* 1968 (8), 102-116.

—, *Psychology as a Human Science: A Phenomenologically Based Approach,* New York, 1970.

Goldstein, Kurt, *The Organism: A Holistic Approach to Biology Derived from Pathological Data in Man,* 2nd ed., Boston, 1963.

Grene, Marjorie, "Straus's Phenomenological Psychology," *The Review of Metaphysics,* 1967 (21), 94-123.

Guilford, J.P., *The Nature of Human Intelligence,* New York, 1967.

Gurwitsch, Aron, *Studies in Phenomenology and Psychology,* Evanston, Ill., 1966.

—, "Edmund Husserl's Conception of Phenomenological Psychology, *The Review of Metaphysics,* 1966 (19), 689-727.

Hamlyn, D.W., *The Psychology of Perception: A Philosophical Examination of Gestalt Theory and Derivative Theories of Perception,* ed., R.F. Holland, London, 1961.

Hampden-Turner, Charles, *Radical Man: The Process of Psycho-Social Development,* Cambridge, Mass., 1970.

Henle, Mary, ed., *Documents of Gestalt Psychology,* Berkeley, 1961.

Howells, William, "Some Present Aspects of Physical Anthropology," *The Annals of The American Academy of Political and Social Science,* January 1965 (357), 127-133.

—, "Supplement: Recent Physical Anthropology," *Ibid.,* May 1970 (389), 116-126.

—. *Mankind in the Making: The Story of Human Evolution,* 2nd ed., New York, 1967.

Jonas, Hans, "Biological Foundations of Individuality," *International Philosophical Quarterly,* 1968 (8), 231-251.

Koffka, Kurt, *The Principles of Gestalt Psychology,* New York, 1935.

Köhler, Wolfgang, *The Place of Value in a World of Facts,* New York, 1938.

—. *Gestalt Psychology: An Introduction to New Concepts in Modern Psychology,* New York, 1947.

—. *The Task of Gestalt Psychology,* Princeton, 1969.

La Barre, Weston, *The Human Animal,* 6th ed., Chicago, 1965.

Lewin, Kurt, *A Dynamic Theory of Personality,* New York, 1935.

—. *Field Theory in Social Science,* London, 1952.

Maslow, Abraham H., *Toward a Psychology of Being,* New York, 1962.

—. "A Theory of Metamotivation: The Biological Rooting of The Value-Life," *The Journal of Humanistic Psychology,* 1967 (7), 93-127.

May, Rollo, Ernest Angel, and Henri F. Ellenberger eds., *Existence: A New Dimension in Psychiatry and Psychology,* New York, 1958.

McGill, V.J., "Behaviorism and Phenomenology: Discussion of A Symposium," *Philosophy and Phenomenological Research,* 1966 (26), 579-588.

Merleau-Ponty, Maurice, *The Phenomenology of Perception,* trsl. Colin Smith, New York, 1962.

—. *The Structure of Behavior,* trsl. Alden Fisher, Boston, 1963.

Rancurello, Antos C., *A Study of Franz Brentano: His Psychological Standpoint and His Significance in The History of Psychology,* New York, 1968.

Rank, Otto, *Will Therapy and Truth and Reality,* trsl. Jessie Taft, 3d ed., New York, 1950.

Rogers, Carl R., *Client-Centered Therapy: Its Current Practice, Implications, and Theory,* Boston, 1951.

—. *On Becoming A Person: A Therapist's View of Psychotherapy*, Boston, 1961.

Scheler, Max, *Man's Place in Nature,* trsl. Hans Meyerhoff, New York, 1961.

Snygg, Donald, "The Need for a Phenomenological System in Psychology," *The Psychological Review*, 1941 (48), 404-424.

Straus, Erwin, *The Primary World of Senses: A Vindication of Sensory Experience,* trsl. Jacob Needleman, New York, 1963.

—. *Phenomenological Psychology: Selected Papers,* ed. Erling Eng, New York, 1966.

—. et al., *Phenomenology: Pure and Applied: The First Lexington Conference,* Pittsburgh, 1964.

—. *Phenomenology of Will and Action: The Second Lexington Conference,* Pittsburgh, 1967.
—. *Phenomenology of Memory: The Third Lexington Conference,* Pittsburgh, 1970.
—. *Aisthesis and Aesthetics: The Fourth Lexington Conference,* Pittsburgh, 1969.
Taft, Jessie, *The Dynamics of Therapy in a Controlled Relationship,* 2nd ed. (by Virginia P. Robinson), New York, 1962.
Van Den Berg, J.H., *The Phenomenological Approach to Psychiatry: An Introduction to Recent Phenomenological Psychopathology,* Springfield, Ill., 1955.

CHAPTER III

THE PSYCHOANALYTIC MODEL

PREDICTION AND CONTROL THROUGH TRAINING OF THE ID

Psychoanalysis, one of the great innovations of the twentieth century, both as a therapeutic technique and as a general personality theory destined to change man's image of man profoundly, begins of course with Sigmund Freud. In a real sense, it also ends with Freud. The great theories which we owe to him as explanations of personality development, deformation, and re-establishment (the key idea of the Oedipus Complex and its psychical adjuncts in personality dynamics, Id. Ego, and Superego) were the result of Freud's self-analysis as much as observational inductions from his therapeutic practice. Insofar, then, as every psychoanalysis undergone in terms of psychoanalytic orthodoxy is a repetition of Freud's self-analysis, we shall be forced to consider the career of Sigmund Freud as the paradigm of psychoanalysis, and only secondarily to reflect upon the "Post-Freudians" and their "Neo-Freudian" endeavors.

There is a practical reason for this procedure. The history of psychoanalytic theory and practice is very complicated from a certain point of view, and exceedingly simple from another. On the one hand, although little can be said of the actual practice of contemporary orthodox psychoanalysts, their theory has not gone beyond the legacy of Freud himself. In the "Post-Freudian" theoretical era, on the other hand, a rather sharp line has been drawn between this "classical" or "orthodox" psychoanalysis and "neo-Freudian" or "revisionist" theories. When one looks more closely at the latter, one finds remarkable divergences, and is soon at a loss about what to call whom or even whom to accuse of what. For example, consider the position of Herbert Marcuse, whose *Eros and Civilization* was an attempt to explicitate the social character of psychoanalysis. He points an accusing finger at the group he calls "revisionists." These include, according to him, Erich Fromm, Karen Horney, and Harry Stack Sullivan, together with Clara Thompson, their "historian."

What he and others reproach to the "revisionists"[1] is not only an outrageous eclecticism but more seriously the unjustified optimism embedded in their consideration of cultural and environmental factors in the shaping of the individual drama. The "revisionists," he claims, are simply combining psychoanalysis, purged of Freud's basic concept of repression, with an idealist ethics. Such a sweeping condemnation by the author of *Eros and Civilization* of those who would introduce "cultural and environmental factors" into psychoanalysis (Marcuse's own career has been characterized by its opposition to the repressions of culture and society) is only the beginning of astonishment. In spite of Marcuse's monumental attack upon Fromm's introduction of an "idealist ethics" into psychoanalysis, Ernest Mandel calls *both* Fromm *and* Marcuse "ethical anthropologists." The main problem, one will surmise, is a lack of space in which to do justice to the positions and claims of warring "neo-Freudians," whether "revisionist" or "culturalist."

A minor practical problem is the following: The "revisionist" and/or "culturalist" neo-Freudians are, in fact, precursers of what we have characterized in another chapter as a "Holistic" movement in the construction of a Model of Man. Indeed, the first "neo-Freudians" were contemporaries and colleagues of Freud himself, as we shall summarily point out. Early minor infidelities to Freud's basic insights were enough to modify the movement profoundly. This practical problem will become, finally, the focal point of the basic problem of this chapter: To what extent has revision glossed over the real message of Freud, and in the last analysis, is the real message of Freud worthy of permanent consideration?

Young man Freud was trained in medicine in Vienna, receiving his medical degree in 1880. The modest means of his parents made it necessary for him to be self-supporting as soon as possible. His university researches had oriented him toward physiology, since Brücke, one of the anti-vitalist oath-takers, was very influential in the university and very friendly to Freud. In 1882, resigned to a medical career when he would have preferred to continue a life of research reserved at the time for the rich), he joined the staff of general medicine of the Vienna hospital for three years, and was soon attached to the psychiatric service of Meynert, where neurology became his speciality. We probably owe to Freud, in this period, the discovery of the

[1] Lacanians call them "The Culturalists" because Lacan is convinced that society can do nothing to overcome the primordial Oedipal situation in which everyone continually lives in no matter what cultural or social circumstances.

analgesic properties of cocaine, although it was Koller who used it effectively as a local anaesthesia and won the palm for the true "discovery."

In 1885, with the recommendation of Brücke, Freud became lecturer in neuropathology at the University of Vienna. Neither this post nor his part time job at a private clinic in the suburbs added much to his store of knowledge nor the advancement of his career; but the same year he was awarded a small travel scholarship, and left for Paris to study with Charcot, the great expert in the treatment of hysteria, and, on his return, spent some time with Bernheim in Nancy.

Back in Vienna, he opened his own practice (as specialist in "nervous disorders") on Easter Sunday, 1886. He soon found himself with a therapeutic arsenal learned in the university but of little help to patients afflicted with "nervous diseases." The classical treatments were electrotherapy (which very soon revealed its inefficacy), and hypnosis—itself held in suspicion by the medical world of Vienna. He had seen hysterical symptoms created by Charcot under hypnosis and abolished by the same means; he had witnessed the therapeutic success of simple suggestion by Bernheim. He read a report of his learnings to the Society of Viennese Doctors, but it was badly received; we have all heard of the derision of the noble Viennese medical profession, and their judgment that the malady called "hysteria" could not be applicable to men, who have no uterus.

In his medical practice, he was associated with Joseph Breuer, fourteen years his senior, who had been using hypnosis, quietly, to cure hysterical patients—the famous "talking cure." Breuer had already spoken to Freud of the case of "Anna O" in 1882; and Freud meditated upon it in the intermediate years, even speaking of it to Charcot. Thus, in desperation, it seems, Freud turned to the one means which seemed to offer hope of results, hypnosis, which Breuer himself seemed strangely less and less inclined to utilize as the years passed. However, with the reluctant cooperation of Breuer, they continued to work together, and in 1885 published a book which has appeared in English under the title *Studies in Hysteria.* Soon afterwards, Breuer parted company with Freud; the publication of the case of "Anna O," which had so much disturbed him before 1882, and continuing recurrences of the same phenomenon—his patients seemed always to fall in love with him—together with the bad reception which the book, though sexually attenuated, received in Vienna, completely discouraged him. The phenomenon of *transference,* and its therapeutic importance, had been discovered, in the wake of defense, resistance, and repression as key concepts. After the departure of Breuer, Freud had no scruples about affirming categorically the sexual etiology of hysteria, and what he was to call the "psychoneuroses." He began immediately to speak of his therapy as psychoanalysis, a method not created from medical theory but structured according, precisely, to the unforseeable progress of the cure, and with the intelligent collaboration of the patient.

Abandoning hypnosis, but keeping the idea of the "talking cure," Freud hit upon the idea of "free association." Free association induced a dream-like semi-hypnotic state in the patient; freed from the "repression" of his subconscious, secure from condemnation by the therapist, the patient benefited.

But mere babbling would not do; the results of hypnosis had shown the harmful effects of repressed memories, and the revelation of these effects in therapy began to show a pattern. The pattern was solidified when Freud's self-analysis discovered that *dreams* revealed suppressed childhood memories and could be interpreted as mythological distortions of suppressed desires which the vigilant *censor* of daylight experience prevented from erupting. Freud understood, then, that the dream is the accomplishment of desire; but not just any desire. Basically, profoundly, it is the desire of the infant to "murder" his father and to "marry" his mother—the primaeval desire of Oedipus, the desire of Freud himself, and (as he was to insist in his metapsychological works) the contradictory desire of humankind to destroy and to unite. This kind of desire, however, is not apparent in the memory we have of our dreaming; consequently, Freud found that there is a *manifest* content in dreams which can be explained, but is not the explanation of the dream. The manifest content is simply the displacement of its latent, and true, content. Henceforth, interpretation, beginning with the *Interpretation of Dreams,* was to be Freud's chief interest in therapy as it was in his theoretical writings. It was to become the focal problem of subsequent studies in psychological theory.[2]

After 1900, Freud began to gather the original group of disciples whom we know so well: Adler, Rank, Sachs, Jung, Ferenczi, Jones.

[2] The whole question of interpretation, or *hermeneutic,* has blossomed since Freud's realization of its applicability in fields other than bible studies. It has become profound philosophical problem. See Richard E. Palmer, *Hermeneutics: Interpretation Theory in Schleiermacher, Dilthey, Heidegger, and Gadamer,* Evanston, 1969.

Paul Ricoeur has recently published a collection of essays, *Le conflit des interprétations: essais d'herméneutique,* Paris, 1969. This book, which will surely be translated soon, should be read in conjunction with this recent *Freud and Philosophy,* New York, 1970.

Among the first psychoanalysts, the "Big Four" who have been the main contributors to the Psychoanalytic Model of Man were Freud, Adler, Rank and Jung; but even before the First World War the group had already broken up. Although, as Progoff so tellingly demonstrated, the contributions of Adler, Rank, and Jung were "integral phases of a cumulative unfoldment of knowledge," they themselves did not clearly understand their own complementarity. They simply approached a common subject from the special point of view of each.[3] Adler was the first to go. His career was early leading him into non-Freudian paths, and his discovery of the "inferiority complex," with its basis in "organ inferiority" and its social consequences and context, was at variance with the Freudian diagnosis of sexual guiltiness as the root of psychic trouble. Adler's defection should not be surprising. If the past weighs so heavily upon the psyche as psychoanalytic theory contended, Adler's past and Freud's past had little in common; nor did their respective "presents." How little resemblance between the sickly little Alfred Adler and the cherished Sigmund Freud of whom maternal affection predicted nothing but greatness. Adler, in his adult years, was a member of the Socialist party and had his medical practice in the poorer sections of Vienna, while Freud lived and thought as a petit bourgeois, with a practice exclusively with the rich middle-class women of the city. Is it any wonder that differing bases of personal and medical experience should lead to differing theories? Indeed, one of Progoff's capital points is precisely the influence of differing personal characters and styles of life upon the formation of the psychoanalytic Model of Man.

Adler finally left the group in 1911. Jung broke with them in 1913, as the result of what were thought to be profoundly differing views between himself and Freud, although Progoff contends that it was Jung who most faithfully exploited the notion of the "depth nature, the psychic magnitude" of the unconscious. The unpleasant and seemingly unavoidable antagonism between Jung and Freud which led to their mutual bitterness on the occasion of the group's second Congress was a near catastrophe for both: Freud had hoped that the support of a respected non-Jewish, non-Viennese medical man would recommend psychoanalysis beyond the narrow frontiers of Jewish Viennese society, while Jung had given up classical medicine in order to be allied with the

[3] Ira Progoff, *The Death and Rebirth of Psychology: An Integrative Evaluation of Freud, Adler, Jung and Rank and the Impact of their Culminating Insights on Modern Man*, New York, 1956. It is only fair to note here that Progoff's thesis is that the final result of this cumulative endeavor is a radically different view of man than the one envisaged in the starting-point.

psychoanalytic group and found himself thrown upon his own resources when the break came. Both men had eventually to come to terms with their personal problems and with each other.

Otto Rank had been chosen by Freud as the chief disciple, and was faithful until 1925, when he published his book on the *Trauma of Birth,* replacing the standard Freudian traumas as explanatory principles of the psychoneuroses with others of his own devising. Rank had attempted to shorten the process of analysis by setting limit dates in advance. (Freud claimed that the dates were really set to coincide with the departure of Rank's vacation ship.) He found that as the time of parting approached, the patient began symbolically to reexperience the trauma of birth, umbilical separation, and new life. Perhaps this situation was reenacted in the life of Rank himself and his relationship to Freud; although Freud publicly praised Rank's work, the psychoanalytic group was quite bitter about its implications, and Rank's departure from their midst was inevitable. Again, Rank's defection from the Freudian group was a loss and a gain simultaneously. Freud lost his most talented disciple, the only one who could match Jung's ability to deal with philosophical, ethnological, and cultural matters; but Rank's departure freed at least part of the psychoanalytic movement from its alliance with the medical profession. Perhaps one of the greatest tragedies of the psychoanalytic movement was the fact of Freud's being a medical doctor; from the beginning, he tought in terms of "pathology." Mental sickness and mental health were primary vectors in the analytic enterprise. Adler, Jung, and Rank, as operators of the "rebirth" of psychology made it possible for psychoanalysis to integrate historical, cultural, social and biological factors. Whether or not this was "a good thing" will have to be considered below.[4]

Abraham, who, with Eitington, had replaced Adler and Jung in the original psychoanalytic group, died in 1925, the year of Rank's defection. Sachs died only in 1947; the others were dead in the meantime (Rank in 1939), and only Ernest Jones was left to write the complete story of Freud's life and work, and he died in 1958.

These men, collectively, left behind them a "dynamic psychology," based upon the idea, mechanistic in origin (although Adler, Jung, and

[4] Rank's American translator, Jessie Taft, was an early social worker. She was the founder of the "Philadelphia School" of social work, but more significant here is her introduction of Rank to Carl Rogers. The influence of Rank upon the early thinking of Rogers was perhaps the most important factor in the formation of the contemporary "Holistic" model of Man, as we have considered it above.

Rank eventually repudiated Freud's materialist concept of the human personality), of the conservation and transformation of energy. Its patriarchal starting-point was, ultimately Leibniz's Monadology, but its life was given scientific stature by Herbart and Fechner. Freud, while he was a student of Brentano, had been urged by the latter to translate one of the works of John Stuart Mill thus coming into contact with the *motivational* aspects of human life which were to be his chief concern, as against the descriptive character of the human "fact" prevalent in mainline psychology since the time of Wundt.

But where is the energy which is conserved and transformed, and where is the motivational principle to be found? Freud found it, through his experience with Charcot, Janet and Breuer, and in his own practice, in the "unconscious."

The idea of the unconscious was not new,[5] but Freud's mechanistic account of it is surprisingly manikin-like in character. Unconscious ideas, desires, functions, successes and failures were seen to be similar in nature to conscious ones, and the reader has the impression of an "unconscious man" who moves and acts within the conscious one. Contemporary, and very orthodox, psychoanalysts, with very little theoretical or philosophical insight, tend to give the picture of man as completely unconscious; his consciousness is only a tricky and unreliable epiphenomenon. Thus, orthodox psychoanalysis is another panel of the psychological dyptich; Wundt and his followers in behaviorism and operationalism rejoin Freud and his followers in orthodox psychoanalysis in their affirmation of our more or less complete determinism—for one group, we are totally determined by society (over which we have no control), and for the other groups we are totally determined by a personal past (over which we have even less control). It should be noted that the orthodox psychoanalyst cannot be wrong; if one agrees with him, all is well; if not, one is manifesting "resistance;" in either case, one is predetermined, in one way or another, to be a "Freudian."

What was the immediate source of Freud's great theories? We have already remarked in passing that it was his own practice, first of all, with its passage from hypnosis, to free association, to the interpretation of dreams which furnished the therapeutic key to the system, as it furnished the key theoretical idea of the system. How did this come about? Freud's father died in 1896, at a very advanced age. It was not an unexpected death, therefore, and its effects upon Freud were for that reason all the more surprising. It provoked a serious mental crisis in

[5] Lancelot Law Whyte, *The Unconscious Before Freud*, New York, 1960.

Freud himself. There were also contributing factors: First, thanks to a long and fullsome friendship with Wilhelm Fliess, of which we have practically a complete record in Freud's published correspondence, he had become sentimentally and intellectually dependent upon Fliess. Second, he was increasingly disturbed about his psychoanalytic practice, with its seemingly insoluble failures, its inexplicable successes, its unceasing ups and downs. In a state of total depression, he turned to practice upon himself the psychoanalysis which he himself had invented. Until then he had not, he admitted, understood at all the process of repression, nor what forces were involved in it; he had come to the disturbing discovery, however, that his patients, who almost universally reported memories of childhood seduction, were simply verbalizing fantasies—imagined by themselves, or induced to imagine them by Freud himself, who had believed, since the publication of *Studies in Hysteria*, that such was the case in all "psychic trouble."

In analyzing his own dreams and his memories and reveries, of which we have an almost daily account in his letters to Fliess from 1896 on, he found the key to the secret drama of every childhood: the Oedipus Complex. He had had a presentiment of this in his practice; he now had the personal experience of it by a courageous descent into the dynamics of his own being. He undertook to relive his own past back to his earliest infancy. Not knowing consciously what awaited him at the end of the road, he nevertheless took the path toward it. What he found was the great drama and the great scandal of mankind. The child which he discovered within himself, with his hostile sentiments towards his father, his incestuous tenderness for his mother, his desire for death, his own fundamental barbarity, became the model of all men. The explanation of the fantasies of his patients with regard to events that had never taken place was that the oedipal situation is lived symbolically in most cases, but no less really for all that. The psychoneuroses had revealed the imporance of dreams, and dreams, in turn, had given the key to the neuroses. His *Psychopathology of Everyday Life* (published in 1904) completed the unity between the strictly pathological and the psychology of "normal people." For the first time, then, Freud was able to universalize his theory by tracing *all* psychic life to its common roots.

This was an ambitious program, and it will perhaps not be out of place to reflect briefly upon some of the factors which determined the kind of universalization Freud eventually carried out. First, as Progoff remarks, the *fin de siècle* transformation of European social and religious contexts and relationships created the symptoms which Freud's patients brought to the analyst's couch: basically a constellation of family and sexual disturbances (rooted in the liberation of women from the confines of the family and the family from the confines of the old class system). He was quite apt to "treat" the "symptoms" which his "patients" presented. Moreover, Freud himself was a striking example of turn-of-the-century European familial dislocation, if only in that he was closer in age to his adoring mother than his father was, for example. Second, Freud himself was a participant in the historic foundational crisis of Western science, religion, and life, in which Science and Ratonality were taking the place of the traditional explanatory function of religious belief. He was apt to share the same faith in the rational interpretation of perennial life-problems; indeed, the very rationality of psychoanalysis was the one thing Freud never questioned, this rationality consisting in tracing "neuroses" to their antecedent causes. His faith in the curative effect of conscious causal analysis was never shaken. Third, Freud was a Jew; and, as David Bakan has pointed out, there are very strong resemblances between the Jewish mystical tradition and psychoanalysis both in method (the master-disciple relationship of the didactic analysis, for example, or the rhetorical covertness of the communication) and in content (that dreams have significance, for example, that human beings are fundamentally bisexual, and so on).

We have become accustomed to suspect theoretical universalizations on the basis of personal experience. We are even suspicious of attempts to find the worker in his work, the artist in his art. Works of art, and even structures, texts, poems, and the like are expected to "stand on their own." How can we reverse our way of looking at "things in themselves" independently of their authors? How can we accept Freud's universalization of his own historical and personal experiencing as valid? One answer (and it should suffice) is that to see the writer in the written, the artist in the work of art, the worker in his work, is not the same as re-creating a biography. We do not necesarily expect the novels of Kafka, for example, to be autobiographical, although we can see Kafka's experiences on every page. We even accept his account of experiencings as making ultimate statements about our fundamental human condition. The same can be said of Freud: If the elements of Freudian psychoanalysis are strongly marked by the life-experiences of Sigmund Freud, we can still accept psychoanalysis as making a statement about the human condition, the statement being made in a thoroughly personal way. This both explains the consistency of classical psychoanalysis and the consistency of neo-Freudian rewritings.

The publication of *The Interpretation of Dreams* (1900) had been received by the reading public with a certain amount of good humor; it was like a book of bizarre old wives' tales. That of *The Psychopathology of Everyday Life* was almost favorable. The publication of a summary of Freud's psychoanalytic theory in 1905, *Three Essays on Sexuality,* was a scandal. It was not, as Marthe Robert remarks, the subject-matter which provoked such general indignation; it was the treatment of sexuality which abolished the frontiers between sexual normality and perversity, and, even more scandalously, the frontiers between adult sexuality and the supposed sexual "innocence" of childhood. Freud was accused of soiling the sacred purity of infancy, of calumniating man himself by replacing the happy innocence of his beginnings on earth with a sombre picture of infantile love and hatred, the child as "polymorphous pervert."

The traditional notion of human sexuality is that it makes its appearance at the time of puberty, with its integrity and its norms already given. Against this idea Freud proposed one of an original sexual instinct in infants which sought satisfaction in some way or another, and which was destined to pass through all kinds of intermediary stages marked by successes and failures and tentative gropings before arriving at the stage of reproduction. The sexual hypocrisy of Victorian Vienna surely played its role in assigning Freud the role of monster; but at this distance it seems surprising that the reading public was not more sensitive to Freud's description of the heroic struggle for maturity and health in which the human infant is engaged from birth, if not before. The human infant is weaker and less able to fight than the animal which he resembles at the same stage; yet he must confront problems which he is liable not to be able to resolve, given the unfinished character of his physical organization and the tenuous means at his disposal.

Freud's concept of sexuality, then, supposed a considerable enlargement of the very idea of sex, once the theory liberated it from the narrow relationship with the genital organs. Thus, human sexuality could be seen as a bodily function which embraced the *totality* of the human being. Thus, too, sexual facts (regarding infants or perverts) could be classified as "normal" or "perverse" not acording to a predetermined norm of behavior, but in relation to the evolution of the *libido*, the unconscious "pleasure principle," which is exemplified at every step, as a success or a permanent or temporary failure. Freud's purpose, then, was not so much to define sexuality, but rather to show, by means of his study of perversions and neuroses that the common notion of sexuality can be maintained *only* by expurgating from it the innumerable *aberra-*

tions which accompany its development. Thus, he studied "normal" sexuality from the vantage point, precisely, of its frequent aberrations. The latter indicate that, by its very nature, the sexual instinct is not necessarily related to an end or object which could define it as a norm.

The first essay, then, on "sexual aberrations" classified with regard to their relationship to a partner (an "erotic object") or a purpose, attempts simply to show that, first, they proceed from a common human bisexuality, and second, that it is curiously the most repugnant sexual perversions which best show the idealization of the sexual tendency of man. The only difference between the pervert and the neurotic is that the former satisfies his desires in reality, whereas the latter, dominated by false modesty, self-disgust, or a sentiment of culpability, represses his desires and lives them only in thought.

The second, and principal, essay, on infantile sexuality, attempts to indicate that the development of the sexual instinct seems to go unperceived and misunderstood as the result of a radical forgetting of man's earliest years. With the forgottenness of infancy are also forgotten the violent emotions which, though decisive for human growth, are thus easily overlooked by parents and educators. In fact, according to Freud, from the very beginning of life, the infant feels sexual desire; his instincts develop progressively until the age of four years, after which development seems to stop completely until the age of eleven. Between the ages of four and eleven, then, the child is in what Freud at the suggestion of Fliess called the period of *latency*. It is during this latency period that the process of *sublimation* occurs, during which the sexual instincts (now without any special purpose) can be turned to all sorts of creative and socially useful activities. Perversion, neurosis, sublimation, and sexual normality are possible end-points of the evolution of the libido, naturally submissive to all kinds of risks.

Freud established the relationship between the perversions and infantile sexuality through the concept of instinct (*Trieb*). The distinction between the various deployments of instinct is established in terms of its source and its end. The source is always some corporeal excitation (erogenous zones), and the end is always the appeasement of this excitation. In the newborn infant, the erogenous zones include, in one way or another, the entire body; but the alimentary orifices and the genital organs are the most sensitive.

The genital organs of infants are not developed enough to play a marked role in the precocious manifestations of infantile sexuality. Freud used the term "sexuality" to cover all the stages of the development of the libido, and understood "genitality" especially to express the final stage of libidinal development. His critics, misunderstanding, read "genitality" throughout. Freud meant to say that before it could be at the service of biological reproduction, the libido expresses and accomplishes itself through all kinds of non-genital functions. Hence, his description of the oral stage of infantile development in terms of sexual accomplishment makes sense if one sees in the libido theory a reformulation of the theory of the conservation and transformation of energy which lies behind it as theoretical background. The oral stage is so-called because of the preponderant role of the mucous membranes of the lips and mouth of the infant in the activity of survival.

Following upon the oral stage of organization, the anal-sadistic phase follows: anal because the erogenous zone becomes the other alimentary orifice, and sadistic because this phase is related to the aggressive behavior of the infant. The anal phase of human development follows the same pattern as the oral phase: its source is an essential physiological function, its end is determined by the activity of an erogenous zone, its sexual valence is *auto*-erotic, since it has to do with no sexual "object."

The properly genital activity of the infant, given the immaturity of the sexual organs, is marked by phases of masturbatory auto-eroticism; the first phase corresponds with the period of breast or bottle sucking, the second corresponds to the brief period of genital activity around the age of four, the third corresponds to the age of puberty. What seems important, however, is that often between the ages of two and five years the infant experiences an erotic attachment to some older person which, despite the uncertain character of "object choice" at this time, is determinant of future sexual-genital development. It is here that real adult seduction most marks the future of human evolution, and it is here that later fantasy "recalls" a seduction-event as having marked the later development of the person for better or worse.

In general, during the earliest years of life, the infant finds in his own body the source of appeasement and pleasure; and his need of satisfaction recognizes neither law nor constraint. He gives every sign of an innate disposition for perversion. Freud's definition of the infant as a "polymorphous pervert" marks, however, the decisive difference between the lawless infant and the perverted adult; the child is disposed for all kinds of satisfactions (hopefully those which succeed in developing along the lines of "normality") whereas the adult pervert is definitively fixed at one or other stage. In both cases, the libido is at work. In the former instance, the child is polymorphously ready to proceed haltingly and hesitatingly on life's way; in the latter instance, the adult's perversion can only expand laterally.

At every turn the human being is faced with enormous tasks. Each time he must adapt to a new stage, the child must renounce the satisfaction and happiness of the former; once weaned from the maternal breast (as he was once cut live away from the maternal body) he must learn "to be clean," and to renounce the intense satisfactions which the products of his own body offer him; he must learn to substitute modesty and shame for his earlier love of what is digusting but free; later, his basic human relationships are upset by the "oedipus complex," in which he lives his first love and his first hate, and he is alone to resolve his primordial conflict. He cannot mature except at the cost of renouncing little by little all the "objects" which he loves, including his own body, nor can he attain full intellectual development except by detaching curiosity from his own sexuality, the source of all his curiosities, and bringing it to bear upon the objects of the world around him. Puberty is, for the growing human being, only another stage of renunciation; it puts him, temporarily, back into the erotic sphere of his infancy, but the prohibition against murder and incest obliges him to another renunciation, to reach out toward another person, finally, with dedication and responsibility.

The *Three Essays on the Theory of Sexuality*, which are so sketchily outlined here, are the basic texts of Freud's theory of personality, as his *Interpretation of Dreams* is the basic text of his theory of therapy. The three essays were continually revised, augmented, corrected, and the ideas contained in them were used as the basis of other works such as his *Introduction to Psychoanalysis* and his autobiography. The reader is enormously aided by the inexpensive paperback edition of the three essays by James Strachey, who painstakingly notes the corrections Freud made in his text in the course of the years, referring the corrections to the works which Freud was producing at the time of the amendment. At the same time, Strachey's notes, especially relating to Freud's correspondence with Fliess, help the student to reconstruct the genesis of the book itself up to the date of its publication.

The theory of the libido, then, was to be the basic notion of psychoanalytic reflection; but in the course of years it had to be adapted and refined in order to meet the explanatory needs of new discoveries in the area of therapeutic practice. Thus, in 1914, Freud introduced the notion of *narcissism*. He himself had wanted to keep the notion of libido essentially linked with sexual energy, while Adler, as we have seen, considered it as a social force and Jung extended it to the point where it became the diffuse and generalized "vital energy" of all nature. The problem was that Jung's practice with schizophrenics and other psychotics revealed that the libido became disengaged from *all* exterior objects to the extent that it could no longer be given a specific valence of any kind; in all other cases the fundamental opposition between the Ego-instincts of self-preservation and the sexual instincts, pushed by the fundamental libidinal energy towards its own goals, could be maintained. The choice which Freud had, then, was either to reduce the libido to an undifferentiated psychic energy (with Jung) or to admit that the Ego itself was endowed with sexual energies and goals. Freud chose a theoretical compromise, not the first nor the last of his carreer, and preferred to admit an opposition not between the Ego and its sexuality, but in the libido itself, qualified by a dialectical opposition: narcissistic libido in the case in which its adherence is exclusively personal, objectal libido in the case in which its attachment is to an exterior object.

The compromise was not entirely successful. Although Freud was ready to admit a "libidinal Ego," he still wished to keep the basic antinomy between sexuality and the Ego. The theoretical problem created anxiety and panic for the psychoanalytic group, which had made a pact not to publish any article which disagreed with the fundamental concepts of psychoanalysis without full discussion with the others. Freud himself explained his intentions only in 1917, in his *Introduction to Psychoanalysis*. He explained that the Ego has, in fact, a double tendency; one is narcissistic and sexual, the other is egoistic and tends toward self-preservation. The two goals might, in any given case, either coincide or diverge. From the narcissistic tendency the Ego-ideal is derived—an interior image which the Ego attempts to realize in order to reestablish the same kind of self-contentment which it enjoyed in the period of narcissistic infancy.

However, another reformulation became necessary, and with it a further complication of the theory, when the observation of children's games, no less than the observation of patients in therapy, revealed a constant tendency toward repetition—an elementary and impulsive tendency which was actualized without regard to, or even in opposition to, the tendency toward pleasurable satisfaction which had up to that time been considered the *sole* principle at work in the dynamic of sickness and health. The pleasure principle, we have noted, was based on the law of the conservation and transformation of energy elaborated especially by Fechner. In the repetition tendency, however, Freud saw the danger in his acceptance of Fechner's law of stability; in repetition, there is the danger, as in animal behavior, for example, of stereotyped automatism. An extension of his phenomenon into the field of psychoanalysis would eventually render all therapy useless. Consequently, in his book of 1920, *Beyond the Pleasure Principle*, Freud had a choice to make: either abandon the fundamental principle of constancy, or to incorporate it somehow into a theory which had always hoped to explain the development of the psychic life. He chose, it seems, the principle of constancy, and asserted that there is a tendency in all living things to seek the stability of an anterior stage of life: death. The non-living stage, in every case, precedes that of life; consequently the end and purpose of life is death, and life itself is nothing but a more or less long or more or less short detour in the universal march toward nothingness.

Within one and the same person, then, according to Freud, *Eros* (the pleasure principle) and *Thanatos* (the tendency toward death) wage a constant confused and indecisive dialectical war in which *Thanatos* tends to reduce all organic life to inanition and *Eros* wages an enduring battle to keep organic life organized. Regarding this speculation, Marthe Robert notes that it was Freud's hope to reduce psychoanalysis to experimental biology, the only "science" which Freud ever really respected. She notes, equally, that the attempt failed, and that Freud could, from then on, only hope that metaphysics could explain what his biology failed to account for.

Hans Jonas, in another context, has a deeper explanation of the failure of this kind of attempt. Ancient peoples, he notes, had the unassailable conviction that *everything was alive* (Heidegger's study of the Greek concept of *phusis*, as reported by Richardson, for example, is an excellent illustration of this); consequently, the primary problem was to explain the possibility of death, hence the ancient and enduring discussions of immortality, transmigration of souls, and so forth. Today, however, in our technological world of "objectivity," with its inheritance of the mechanistic models of intervening centuries, we must proceed as though *everything is dead*; consequently our primary problem is to try to understand the possibility of life, hence our contemporary curiosity about the origins of life on earth in the sciences of paleontology, archeology, and so on, and our preoccupation with the sources of life in living beings (biochemical research on the DNA molecule, for example). The entire contemporary enterprise—and Freud is our contemporary here— is the heritage of Nietzsche who saw that "*God* is dead." The loss of eternity in the search for a dead present "is the disappearance of the world of ideas and ideals in which Heidegger sees the true meaning of Nietzsche's

'God is dead'." [6] The very title of Freud's book, *Beyond the Pleasure Principle,* is suggestively related to the thought of Nietzsche.

This, of course, is only theoretical. Whatever the psychoanalyst may think of the rigorous conclusions drawn by Freud, he cannot make use of the concept of a death instinct as a therapeutic instrument, he cannot accept a *fundamental* tendency in the human being toward destruction and self-destruction, a fundamental sado-masochism; he can, if he wants to "cure the patient," count only upon *Eros.*

Freud himself, however, continued upon his way, drawing the sad conclusions from his theoretical reasoning applied to biology. The main result was what today is popularly considered his main contribution: the concept of *The Ego and the Id* (1923) a late elaboration of his libidinal theory imposed by increasing clinical data which had to be taken into account. It was becoming clearer and clearer that the Ego was not entirely coincident with conscious existence, nor was it as opposed to libidinal instinct as it was supposed to be. In reality, the Ego was in conflict with a whole constellation of contradictory impulses, both erotic and death-seeking.

What was going on between 1920 and 1923? According to Progoff, the customary general agreement about the "obscurity" of Freud's *The Ego and the Id* is removed if we understand that the book "is best understood when it is read as part of a continuous (though undeclared) dialogue with Jung on the nature of the deep conscious." The reformulations of theory which Freud continually and reluctantly undertook were veiled responses to Jung, the "responses" taking the form of attempts to integrate Jung's unassailable findings into what Freud had hoped would be a coherent theory. The problems which narcissism and death posed for Freud's theoretical reason (as for his own life) could not be overlooked, yet could only be integrated in theory and in life with great difficulty—and an equally great obscurity. In the course of his reformulations, then, Freud, willingly or not, gave up his primordial distinction between conscious and unconscious, while trying (by a series of otherwise incomprehensible distinctions) to keep the original theory of libidinal repression from flying to bits. His final statement can be summarized as follows:

All of the instances of personality organization proceed from the impersonal, obscure, impenetrable *Id*, full of impulses of all kinds but without organization or direction—the chaos of the unconscious life,

[6] Hans Jonas, *The Gnostic Religion: The Message of the Alien God and the Beginnings of Christianity,* 2nd ed., Boston, 1963, 334-340.

which knows neither negation nor contradiction, duration or time, value, good or evil, or morality. These must be the construction of the other personal dynamisms which attempt as well as possible to account for the organization and maintenance of the self.

The Ego, which alone can speak, is alone the seat of *logos*, of reason and order, is merely a function of the Id, insofar as it is influenced by the exterior world.

The Superego is formed through the relationship of the child with its parents, through a process of indentification and introjection. In this process, the external world of authority, of value, of moral precepts, is internalized. Since its origin, like that of the Ego, is basically the impersonal and unconscious Id, the Superego is likewise partly unconscious, although it exercices judgment over the impulsiveness of the Id. Thanks to its archaic origin, the Superego is the personal representative of the past, of tradition, of the non-existent, and its tyranny is a threat to psychic stability. But the Id, with its chaotic impulses towards both life and death, poses an equal threat to the Ego, menaced "from below" by almost uncontrollable and unconscious desires and "from above" by the crushing weight of reality and society. The whole emphasis of Freudian psychoanalytic therapy is Ego reinforcement, since the Ego—even though half-sunk in unconsciousness—must attempt to reduce the exigencies of the Superego to reason and to tame and control the chaos of the Id.

The least enlightened among psychoanalysts (especially Americans) tend to classify these three valences exclusively, and even to reify them. European analysts tend to understand them as various functions of one and the same psychic existence. The American psychoanalytic enterprise is thus truer to the letter of its origins, truer to the mechanistic and deterministic theoretical background of psychoanalysis while, in general, the European enterprise tends towards a holistic and humanistic view —with very notable exceptions. This, in part accounts for the " parallel" International Congress of Psychoanalysis in Rome in The Summer of 1969. Dissident members, protesting against the preponderance of American analysts in the official congress, and against the oligarchic character of a movement which has become, in addition, a gerontocracy, held a "Counter-Congress," in which they attempted to exhort psychoanalysts to call themselves into question and to enter into contact with the ideological, economic, social, and political realities of the day. The introduction to the "official" congress by Mitscherlich was carried on in classical and orthodox fashion, and was, in his own words, "as lovely and useless as a promenade in the forum." [7] Observers have not failed to note the similarities between the conduct of the International Congress of Psychoanalysts in Rome and the Synod of Catholic Bishops in the same city recently. Grossly put, it is a contest between Church and church, and reporters have not failed to remark upon the "conquest of Rome" dreamed by Freud in his dream of being Hannibal. In a more subtle way, each "church" has its orthodoxies and its heresies, and each, in the same city, has attempted to come to terms with the dissenting currents within it.

It is difficult to say whether the Freudian project is finished. Some maintain that the very disengagement of the concept of Id, Ego and Superego are worthy of a lifetime of patient reflection and practice, and it is not surprising that Freud himself had only time to develop a psychology of the unconscious. After his death, they point out, when the Second World War demonstrated concretely the sad theory of the savageness of the human animal, it was the *Ego* psychology of people like Heinz Hartmann which reestablished a semblance of at least theoretical human dignity. At the present time, O. Hobart Mowrer, on a very slim experimental basis, to be sure, is crusading for the explication of a psychology of the *Superego* which might bring the whole enterprise into balance. Some judgment upon these questions will be attempted at the end of this chapter.

Research has not ended there; and the interested student can follow the psychoanalysis of infants from the earliest efforts of Pfister, Freud's "dear man of God," the publication of the case of "Little Hans" (treated not by Freud but by Hans' own father), and that of "Little Arpad" (treated by Ferenczi). Anna Freud, Sigmund's daughter and now the Grand Old Lady of Psychoanalysis, has spent her entire life in the psychoanalysis of children; and Melanie Klein in England and Louise Despert in the U.S.A. have pushed the theoretical reflections even farther. The interest in infantile analysis is that it confirms, at every step, the grand lines of Freudian theory as it is revealed in the psychoanalysis of adults, giving great suasiveness to the theory.

It is impossible to do justice to the original Freudian theses in these lines; the intention has been simply to guide the reader in his consideration of Freudian theory. Consequently, in addition to the broad sketch of therapy and personality theory, a few words must be said about Freud's metapsychology, however summarily.[8]

[7] Reported in *Le Monde* on July 30, 1969.

[8] This use of the word "metapsychology" is a preference of the author. At least two other meanings are given to the term in psychoanalytic literature, neither with any better reason.

Besides Freud's reports on progress in the technique of therapy and his continual expansion of personality theory, there is a curious series of texts which attempt in some way to extend the implications of Freud's theory to the entire range of human existence. *Totem and Taboo* (1912) is the first of these books. In it, Freud attempts to show the relation between religion and the human psyche—religion being the basic sublimation of the human libido. After a study of the rites and laws of totemic religion, based on the work of Darwin, Atkinson, and Robertson-Smith, Freud found his theory of the oedipus complex confirmed. In the simplest of terms: the totem is considered the father of the tribe; it is forbidden to kill and eat the totemic animal, and to have sexual intercourse with a woman belonging to the totem. Here one sees the renunciations basic to human social existence; murder and incest must be proscribed for the good of society. (Freud never made a moral judgment about the nature of the libido, it should be pointed out; it is not that the desire to murder one's father and to have intercourse with one's mother is either good or bad, the goodness or badness of wish fulfillment being an interdiction laid upon the individual by society as a *consequence* of the disasterous results of living out these desires in reality.) The problem of survival is paramount; internecene war must be restrained and exogamy must be practiced if we are to live together and to survive our own instincts. In *The Future of an Illusion,* Freud attempted to consolidate his position against organized religion as a "force for good" in the world. As men grow more wise, more "scientific," and thus less religious, the ancient taboos proposed by the Church will grow weaker. In order to save society, then, we must recognize, he thought, the weakening of the proscriptive power of religion and the strengthening of reason through science. Thus there is hope that when the "illusions" (i.e. wish-fulfillments) of religion give way before the untameable instincts of men, the last bastion of defense will be reason, whose power has been proved in the modern scientific enterprise. In *Civilization and Its Discontents,* he attempts to come to terms with the basic restlessness of men in society, which must necessarily be oppressive/repressive, as must the illusory Church. Finally, in *Moses and Monotheism,* he attempts to extend his theory to the totality of human history, especially religious history. The earlier parts of the book were published early in Freud's career; they attempt to reason out the *kind* of monotheism which Moses advocated. Moses, he found, was an Egyptian, a noble follower of Amenhotep IV, who had destroyed the cult of the dead and established a kind of cosmic monotheism based on the worship of the solar disc. In order to explain the contradictory attributes of the God of the Hebrews, Freud postulated an amalgamation of the religion of Moses with the worship of Jahveh, a volcano-demon, practiced by the Medianities whom the Hebrews met at the Oasis of Qadeš. The latter group was led by another Moses who, surprisingly, was not an Egyptian. The "real" Moses, according to Freud, was murdered by his followers before they reached the promised land of Canaan; and it was the memory of this primaeval murder of the father of the tribe, guiltily repressed in the memory of Israel, yet breaking out from time to time in prophetic utterance, which, with "the return of the repressed," was the cause of the re-enactment of the original sin: the crucifixion of Jesus. Thus, the thesis of religion as a neurosis was reinforced, and the application of the theory to the totality of human history (with the aid of a postulated Darwinian "primaeval horde" the "sons" of which murdered their "father" and thus set in motion the entire human enterprise) made of Freudian theory a complete explanation of human history and society, in its collectivity as well as in its individuality.

Freud's decision to publish the final sections of *Moses and Monotheism* is touchingly detailed in his own preface to the complete edition, begun in the Spring of 1939 in Vienna, and completed in the Fall of the same year in London, to which he had fled, with the help of Ernest Jones, before the Nazi invasion. He had withheld publication of the final section, he said, as a matter of self-protection with regard to the Catholic Church in Austria. The Church, he felt, was hostile to him, as he was towards the Church, and was bound to react violently to the publication of his critique. At the same time, he realized, it was the Church in Austria which was protecting him, as a Jew, against Nazi persecution. When he realized the church was no longer able to fulfill its promise, he felt free to publish the manuscript.

Freud's critique of religious belief was focussed mainly against the Catholic Church of Vienna (after all, the only one he knew). His condemnation of the Church in *Moses and Monotheism* as "the implacable enemy of the freedom of thought and the progress of knowledge" dates from 1939, but translates an opinion formulated in his earliest years at the university, in 1873. With regard to Viennese Catholic society, which he always identified with the Church, he felt from the beginning the force of his Jewish minority position, and thus his exclusion from the nationality of the others. His liberal political convictions and his scientific activity (and its recurrently hostile reception) added to his own hostility to the Church. Thus, his critique of religious belief in general resulted from his own "natural atheism," reinforced by his psychoanalytic reflections, as we have seen from *Totem and Taboo* in 1912 to the preface and final section of *Moses and Monotheism* in 1939. On the other hand, his condemnation of the Church owes more to the political and philosophical judgment of Freud than it does to psychoanalytic theory, as Daniel Widlocher has pointed out. In fact, Freud's psychoanalytic practice had practically nothing to do with the Catholic community; both patients and followers were recruited within the Jewish community. The earliest non-Jewish psychoanalysts were liberal protestant ministers, Pfister and Keller, for example.

However, the Roman Catholic clergy began to practice psychoanalysis as early as 1925, and the expansion of it was favorized by the emigration of many Austrian psychoanalysts to "Catholic" countries during the 'thirties. After the war, it seemed that psychoanalysis and religion were not as incompatible as Freud had maintained. Religion, as a displacement of libidinal impulses, may be either sublimation or neurosis, depending upon the case. Besides, there was a growing interest on the part of theologians and moralists in psychology and psychopathology. In the present state of affairs, there are divergent and opposing camps. One group of psychoanalysts, medical practitioners and theologians is convinced that no psychoanalyst true to the name can maintain any religious faith, much less continue a Catholic. Another group, of course, maintains the contrary opinion. Widlocher thinks both positions are contestable: some psychoanalysts, in the tradition of Freud himself, work, as did Freud, with a sclerotic and caricatural image of the Church; some Catholics, on the other hand ("according to a process with which they are familiar," remarks Widlocher) amalgamate scientific critique with moral objections.

These few remarks upon the confrontation of psychoanalysis and religious faith are introduced in order to take advantage of the problem which Daniel Widlocher, to whom we are indebted for his analysis of the deeper meanings of the twenty-sixth annual congress of the International Psychoanalytic Association in Rome (27 July - 1 August, 1969) [9] has so well underlined. There is a danger in the penetration of Freudian psychology into religious circles—a danger of equivocation on both sides. For the Catholic clergy, in particular, there is the danger of psychologizing pastoral dialogue in order to mask doctrinal discomfort, of expecting psychoanalysis to smooth over conflicts or to repair the inadequancies of

[9] *Le Monde,* 26 July, 1969.

defective personal commitments. Once upon a time, he notes, psychoanalysis had the reputation of guaranteeing divorce; nowadays we attempt to patch up marriages with it. Since the whole of the Freudian enterprise is the unveiling of hidden meaning, it *can* be reestablished as a phenomenology of human experience and human interrelationships, and thus perhaps its true force denied. The biological and social determinisms which Freud posited as governing the whole of human life, human development, and human relationships, the reductivist thinking which was Freud's explicit work, his intense suspicion of human liberty as epiphenomenal may, in such hands, be simply ignored, and thus Freud's real contribution to the question of man, with its enormous problems, will be lost.

What then, in sum, was Freud's real contribution to the resolution of the question about man? Widlocher's remarks lead us to remember that Freud himself was one of the great pessimists of modern times. (One thinks of his discouraging letter to the mother of a homosexual, of his remark to another who hoped to raise her children in the best psychoanalytic way: "Do as you like; it is sure to turn out badly.") He unveiled the human problem in its profundity and its amplitude, but by that very unveiling gave no hope that it could be forgotten. The best that Freud's own psychoanalyzing could hope for was a kind of reformation, a more or less successful adaptation to conservative social living; orthodox psychoanalysts today receive their patients as little robots, badly conditioned by family and social contacts, and send them away as refashioned robots, re-conditioned by psychoanalysis to participate in what passes for our common life.

Marthe Robert has beautifully summed up the kind of thinking which was Freud's, and we shall paraphrase her summary here: Freud, in spite of his conscious desire for unification in theory, is in fact fundamentally dualist. All of his grand concepts are double and conflictual: conscious-unconscious; Ego instincts-sexual instincts; the pleasure principle-the reality principle; Ego-Id; life instinct-death instinct, this last being the basic antithetic duality which englobes all the other oppositions. His doctrine of the libido, she continues, passes through three phases: in the first, sexuality dominates the picture; in the second, narcissism is primordial; in the third, death triumphs over life. These phases mark also the decisive steps in the life of Freud himself, and to the extent that orthodox psychoanalysis holds explicitly to the letter of Freud's reports, the statement in the beginning of this chapter is reconfirmed.

For Freud himself, man is fundamentally and dialectically good-bad.

Fundamentally, since he is subject to a radical conflict which is the source both of progress in living and of regression, or of a perpetual regressive progress. Dialectically, since his nature is unstable, formed around an ego which itself is unstable, formed in its turn genetically, and continually beseiged by the antagonism of Eros and Thanatos and the permanent struggle between instinct and repression, the Ego-libido and the Super-Ego. Edgar Morin too has well described the Freudian Man: "The most remarkable thing about the axis of Freudian anthropology is that man (good-bad) is constitutionally neurotic-well. Man lives a permanent neurotic situation which is the condition of his health." [10] Again: "While Marx surpasses his pessimism by messianism (the great hope of great pessimists) Freud is mute, since that kind of revolution would suppose a structural and multidimensional transformation the key of which he does not believe he possesses [11] precisely because he has localized the key points. Freud is even very pessimistic about civilization, that minimum policed humanity, of which he knows the extreme fragility. He does not even advocate sexual liberation... He fears, on the contrary, the freedom of instincts. He knows that civilization is necessarily repressive, that it is derived from repression. He knows the terrible forces enchained by the Superego, and is not willing that they be unleashed." [12]

Taking into account, then, Freud's own well-founded pessimism, taking into acount also Widlocher's warning that Freud's own reductionism and determinism may not be forgotten, what can possibly be said?

The first thing to be said is that the Freudian theory, in some circles, continues as before, even though in another key. Our great, mad contemporary, Jacques Lacan, together with his disciples, carries on in his own way a kind of determinism which, in strict concord with the original Freudian theses and in collaboration with new structuralist theories, remains true to the fundamental pessimism of the founder. Lacan's famous article on the phenomenon of the "mirror image" (the sight which the child has of himself in a mirror as a *whole* being is only an abiding phantasm which cannot, inspite of all efforts, be realized concretely) with its explicit reference to the phenomenon of "imprinting" in birds and animals, is a *manifesto* of the Freudian movement which can only be called "Lacanian," since the great man admits no masters except Freud and disowns all discipleship including that of Foucault and Althusser.

[10] Edgar Morin, *Introduction à une politique de l'homme*, Paris 1969, 24.
[11] The Reference is to *Civilization and its Discontents*.
[12] Morin, 27.

A second thing to be said is that while very "orthodox" practitioners of psychoanalysis hold theoretically to all the "orthodox" theses of the founder of psychoanalysis, nothing can be said of their use of it in therapy, since therapeutic process in therapy is not controllable by any means except its continual and consistent confirmation in practice. Whether this confirmation is provoked by the analyst or is evoked in the patient will forever remain a mystery.

A third thing to be said is that, obviously, Freud always surpassed, in his own work and thought, the explicit theses which he advanced. The very effort at "therapy" which has as its goal the possibility of liberation, of responsibility, of love and dedication (the close followers of Freud, in their practice as well as in their theories, had put their hopes entirely in the healing power of love) marks, if only unconsciously and unexplicitly, a negation of Freud's own explicit philosophy. It is on this point that psychoanalytic theory is making progress.

Georges Politzer has shown in a striking way the sharp break which exists between psychoanalysis and "classical" psychology: the latter is abstract, the former concrete. It is a useful classification, yet full of contradictions too. For example, Paul Ricoeur has shown that the key concepts of psychoanalysis are not expressible in operational terms (thus bringing to naught the efforts of the Minnesota laboratories), yet this very fact is food for thought: if psychoanalysis is concrete, if as Politzer stated, "The very way in which the problem of the dream is posed implies a definition of the psychological fact which displaces interest in spiritual entities towards the dramatic life of the individual," [13] and if the total enterprise of psychoanalysis is precisely the analysis of the individual "drama," why cannot its very concreteness be examined? Politzer recognized the inescapable duality between the concreteness of the problems of psychoanalysis and the abstractness of "the notions it uses, or creates, and the schemes which it uses." [14] Politzer himself explained his own use of the term "concrete," and in general how the concrete fact can be found again if one retraces inversely the development of the abstractions. This point is underlined here because the attempt of Politzer to explain, or at least understand, Freud's seeming infidelity to his own principles is the starting-point of contemporary reflection.

A fourth, and final, thing to be said is that it has seemed useless to make a blanket counter-proposal of *absolute indeterminism* against the

[13] Georges Politzer, *Critique des fondements de la psychologie,* 2nd ed., Paris, 1967, 110.
[14] *Ibid.,* 209.

total *determinism* which the desolate theses of Lacan, for example, advance. In the first place, the term "determinism," as we have seen all through this chapter, has valences which mere words cannot do justice to, if the term is taken in a univocal and classical sense. In the second place, the notion of total indeterminism, as it has been applied, for example to "the will" and "human liberty" could not stand up under the weight of an argument. To the question, "Are we completely determined?" the answer is usually, "Are we completely free?" Neither does full justice to the inescapable discoveries of Freud—nor to the reality of our human existence. Classical (and especially neo-Scholastic) philosophy, with its moral and religious stake in freedom and responsibility, has chosen to regard human freedom as an *autonomous* power. It is informed by the intellect, to be sure, but the "power of choice," of "choosing between alternatives," the famous "freedom of indeterminacy," was seen as infringed upon only by *external* pressures, like fear, intimidation, invincible ignorance, and so on. The inestimable contribution of Freud to this doctrine has been the insight that, though we are indeed free, we are not so totally free as to be liberated without effort from the biological, social, and psychological conditions of human existence, which are *internal* conditions of our human freedom. The ability to accept fully the contradictions of Freudian theory (unlike the either/or intransigent options of Lacan and Sartre at each of the spectrum) has been the great contribution of Louis Monden, sketched in his book, *Sin, Liberty, and Law,* in a religious and pastoral context, and by Paul Ricoeur in his *Voluntary and Involuntary* and his *Freud and Philosophy* in a more philosophical context, but with no less concern for the moral and religious character of the question. In the practical order, following upon the theoretical acceptance of the contradictions inherent in a psychoanalytic view of man, there is the possibility of accepting as one's own the restrictions on our individual freedom—whether they are imposed from without or are inescapably part of our own human constitution—by means of a fundamental and free option which sets our course consciously in the direction of human growth and achievement. If "philosophical anthropology" has any task to perform in this area, it is the development of a concept of human freedom and creativity which takes into account not only the imperious "will" which is supposed to achieve its goals in accordance with strict obedience to the "laws of nature," but the very conditions of that nature and that willing.

SELECTED ADDITIONAL READINGS

FOR CHAPTER III

Bakan, David, *Sigmund Freud and The Jewish Mystical Tradition,* New York, 1965.

Becker, Ernest, *Angel in Armor: A Post-Freudian Perspective on the Nature of Man* (dedicated to the Memory of Emmanuel Mounier), New York, 1969.

Breuer, Joseph, and Sigmund Freud, *Studies in Hysteria,* Boston, 1961.

Erikson, Erik, *Childhood and Society,* 2nd ed., New York, 1964.

Fingarette, Herbert, *The Self in Transformation: Psychoanalysis, Philosophy, and the Life of the Spirit,* New York, 1963.

Freud, Sigmund, *The Standard Edition of The Complete Psychological Works of Sigmund Freud,* trsl. under the editorship of James Strachey in collaboration with Anna Freud, 24 vols., London, 1966.

Fromm, Erich, *Sigmund Freud's Mission: An Analysis of His Personality and Influence,* New York, 1963.

Hartmann, Heinz, *Ego Psychology and the Problem of Adaptation,* trsl. David Rapaport, London, 1958.

—. *Essays on Ego Psychology: Selected Problems in Psychoanalytic Theory,* London, 1964.

Homans, Peter, *Theology After Freud: An Interpretive Inquiry,* New York, 1970.

Jones, Ernest, *Sigmund Freud: Life and Work,* 3 vols., London, 1956, 1957.

Jung, Carl G., *Memories, Dreams, Reflections,* ed. Aniela Jaffé, New York, 1961.

Marcuse, Herbert, *Eros and Civilization: A Philosophical Inquiry into Freud,* New York, 1962

Nelson, Benjamin, ed., *Freud and the Twentieth Century,* New York, 1966.

Progoff, Ira, *Depth Psychology and Modern Man: A New View of the Magnitude of Human Personality, its Dimensions and Resources,* New York, 1959.

Rank, Otto, *The Trauma of Birth,* trsl. Jessie Taft, New York, 1929.

Ricoeur, Paul, *The Symbolism of Evil,* trsl. Emerson Buchanan, New York, 1967.

Robert, Marthe, *The Psychoanalytic Revolution: Sigmund Freud's Life and Achievement,* trsl. Kenneth Morgan, New York, 1966.

Robinson, Paul A., *The Freudian Left: Wilhelm Reich, Geza Roheim, Herbert Marcuse,* New York, 1969.

Szasz, Thomas S., *The Myth of Mental Illness: Foundation of a Theory of Personal Conduct,* New York, 1961.

Thompson, Clara, *Psychoanalysis: Evolution and Development,* New York, 1950.

CHAPTER IV

THE SOCIOLOGICAL MODEL

FROM DOING GOOD TO BEING DONE

Sociological science in America is indeed an imposing affair, and even if one has certain misgivings about it one cannot overlook sociology when dealing with "Models of Man." There is no doubt that this science has won a place of esteem and power on these shores in the past hundred years. Since 1928, when Sorokin's early overview of sociology appeared, listing more than a thousand "notable" figures and twelve "schools," [1] the field has expanded without surcease. The latest overview no longer even attempts to keep track of the thousands of books, articles and monographs which have contributed to the field;[2] a much-quoted and highly esteemed introduction to sociology abandons the attempt at systematization entirely, and presents for consideration mainly the thoughts of the "founding fathers" of sociology, Comte, Weber, Durkheim and Spencer.[3] It is not that sociologists have not tried and are not still trying to bring some intelligible order into their disorderly house;[4] it is that they have been unable to find a unifying key. From the most imposing milestone monograph to the humblest introductory text, sociological authors are unanimous in their appraisal that the complexities and ramifications of sociology as a science cannot be in any way systematized. One well-known French expert (Lucien Karpik) has concluded that if the criterion of a real science lies in the accumulation of wisdom, sociology is not even a science, merely an accumulation of discrete data and theory.[5]

If we had hoped that the science of sociology would reveal to us

[1] Pitirim Sorokin, *Contemporary Sociological Theories*, New York, 1928.

[2] Walter L. Wallace, ed., *Sociological Theory: An Introduction*, Chicago, 1969.

[3] Alex Inkeles, *What is Sociology? An Introduction to the Discipline and Profession*, Englewood Cliffs, N.J., 1964.

[4] See, for example, Mitchell Duncan, *A Hundred Years of Sociology: A Concise History of the Major Figures, Ideas and Schools of Sociological Thought*, Chicago, 1968. Duncan identifies at least twelve different schools, few of which coincide with those identified by Wallace—though Duncan has been charged with omissions.

[5] *Le Monde*, September 20, 1967.

something about ourselves, what would that have been? Evidently, we would have expected to learn something, not about a solitary ego in a strange landscape, but about ourselves as social beings—which indeed we conceive ourselves to be. We would have expected to learn something, surely, about the "essence of society," but more important about society as constitutive of ourselves and ourselves as constituted by and constituting society. Yet, when recently the president of the American Sociological Association urged that men be "put back" into sociological consideration he was accused of a "disappointing appeal to a reductionist psychologism." [6]

Let us take the extreme case seriously. Assuming that sociology is not a "science," that is does not contribute to the accumulation of wisdom concerning human-being-in-the-world, why have we been led to accept its prophetic importance in our common life? We shall attempt to find a few major themes in the history of sociology for our purpose, but we shall begin with acceptance of the defeat of our expectations that sociology might provide us with a notion at least of "society at large" if not of "social man." What will account for this failure, if the charge be true? And what has been delivered instead? The following summary paragraphs will attempt to indicate that a model of man which one might have expected from the science of sociology must be inferred from the methods and results of that science, and that the methods and results of sociology reveal more of the being-in-the-world of the sociologist-as-expert than of the common man.

Contemporary American sociology owes little except its inspiration to its European founders and their Continental successors. The former, in general, enjoys a more empirical orientation (the effects of which we shall examine in due course) while the latter have been especially strong in social theory. The hope that empirical studies and theory might be brought closer together, in sociology as in psychology, has been late in fulfillment, and has not yet been entirely successful. Europeans lack American money to fund the expensive empirical research projects, while Americans are leery, as usual, of theory which is not directly applicable. The French demographer, Alfred Sauvy, and the immigrant American sociologist, Paul Lazarsfeld, have enjoyed the greatest success until now in unifying these two opposing tendencies.

[6] See George C. Homans, "Bringing Men Back In," (Presidential Address delivered at the Annual Meeting of the American Sociological Association in Montreal, September 2, 1964), *The American Sociological Review,* 29 (1964), 809-818. For the rejoinder, see Walter Buckley, *Sociology and Modern Systems Theory,* Englewood Cliffs, N.J., 1967, 36.

All sociologists have been inspired by the great visionary August Comte (1798-1857), the inventor of his own brand of "positivism" and of "sociology." Comte's positivistic abolition of the *causal* concept from science (maintaining that it is a philosophico-theological, and not a scientific, idea), together with his search for the true general relations which link "social facts" together were to be important in the endeavor which followed his grand theory. To him we trace a double movement in the search of sociology: first, the desire to find general social laws (thus making sociology scientific), second, the ambition to put these laws to use in social action (thus inviting sociology to occupy the seat of political and social power formerly held by philosophers and theologians).

The work of Comte, then, had largely inspirational value upon the other founding fathers of sociology: the inspiration to make the study of society a science. In France, Emile Durkheim (1858-1917) realized that in order to be a science, sociology had to find some*thing* to study, on the classical model of "real" science. Social facts, therefore, had to be redefined as "objects," comparable to the objects which the impressive physical sciences study. On the basis of the theory also utilized in his famous *Elementary Forms of the Religious Life,* Durkheim defined a "social fact" as "any kind of action, whether of a set nature or not, capable of exercising external constraint over the individual." The other important ingredient of the social fact was that it must be "general throughout a given society." [7] The two inseparable facets of the social fact (its generality throughout a given society and its property of external constraint over the individual) were unified by Durkheim's positing a collective consciousness distinct from the totality of individual consciousnesses in the society, with the result that social facts can be seen as radically distinct from individual psychological facts. The latter are not—at least sociologically—of interest for science; the effort to isolate the former was simply an effort to stabilize social process in such a way that it could be considered scientifically. The problem, Durkheim maintained, is that

> the greater part of our social institutions were bequeathed to us by former generations. We ourselves took no part in their formation. Even where we have collaborated in their genesis, we can only with difficulty obtain even a confused and inexact insight into the true nature of our action and the causes which determined it.[8]

[7] Emile Durkheim, *Rules of the Sociological Method,* 8th ed., trans. Sarah A. Solovay and John H. Meuller, ed. George E.G. Catlin, Glencoe, 1938.

[8] *Ibid.*

Thus, for the social scientist, "the important thing to know is not the way in which a certain thinker individually conceives a certain institution, but the group's conception of it." However, the group's conception of social institutions lies also under the pressure of external constraint; and this, Durkheim claimed, is one of the tests for a truly social fact:

> When I fully conform and assent ... the constraint is felt only slightly, if at all, and is therefore unnecessary... I am not obliged to speak French to my fellow-countrymen, nor use the local currency, but I cannot possibly do otherwise... It becomes immediately evident that all education is a continuous effort to impose on the child ways of seeing, feeling and acting which he could not have arrived at spontaneously... If the constraint in time ceases to be felt, it is because it gradually gives rise to habits and internal tendencies which make the constraint unnecessary.[9]

One can here see the prolongation of Durkheim's efforts to explain *religion* as a social phenomenon on the basis of its generality and its character of external constraint (the constraint not being thought of pejoratively, but as both uplifting the individual and imposing obligations upon him).

The same reasoning will be applied in another chapter of this book [10] to the institution of *language*: Language, as a social fact, can be considered an object distinct from our use of it as individuals. It, like the rest of our social institutions, is entirely inherited from other speakers in our society; it conforms to the requisites of a social fact in its being general throughout a community and its exercising an external constraint upon those who use it. Language follows the patterns of the social fact even more closely, since although we may rebel against the institutions of our society, and even seek to overthrow the total social fabric by revolution, we cannot escape the constraints of our language if we wish to communicate through it. For those who use the language they have learned, those who have mastered it, there is no feeling of "constraint."

An important feature of language as a social fact in the present context is that, in Durkheim's view, it must be considered *independent of historical development*. Durkheim did not consider recent societies simple duplications of their predecessors, however; "a group which succeeds another is not simply a prolongation of the latter ... it is qualitatively different from it, having gained some properties and lost others. It consists of a new individuality." If, then, each generation is new, how deal with the original question, that is, how to explain society and/or language as a

[9] *Ibid.*
[10] See Chapter VI.

fixed fact in the scientific sense? Durkheim answered that "the peculiarity of social constraint is that it is due, not to the rigidity of certain molecular arrangements, but to the *prestige* with which certain representations are invested." Consequently, attention must be paid primordially to the general, objective characteristics of language as a social fact, with minimal attention to the performance of any individual speaker. The question of "the prestige of certain representations" must be kept in mind for later exploitation, since one of the problems of structuralism [11] is that it must proceed *as if* nothing changed in society, in language, etc.), while taking into account that, as a matter of fact, these *do* change, such that the "new individuality" of each generation recognized by Durkheim can also be brought under scientific scrutiny.

Durkheim may be considered the founder of empirical studies in sociology in the tradition of Comtean positivism, as well as a theoretician. His impressive field work studies in anthropology gave a solid base to his theory of the foundational character of religion in society; and his study of the phenomenon of suicide was inspired by the same kind of opposition to "psychologism" which exercised Brentano and Husserl: the theory that the laws of psychology are the laws of logic, science, and sociology. Durkheim attempted, in opposition to psychologism, to prove that the phenomenon of suicide, for example, is explicable by *social* factors as well as by individual psychological causes. However, with his typically Gallic flair for overgeneralization, he ended by producing what is, in the last analysis, a theory of society in which men are wholly determined by society.

Marcel Mauss (1872-1950), a student of Durkheim, abandoned the evolutionary and quantitative typology of his master, and hoped rather to elaborate a qualitative typology of social phenomena so as to distinguish sociology from history in a more effective manner. He devoted his energies to the holistic view of society, following the theoretical stance of Durkheim, and considered, as against the partial field studies of his teacher, *total* social phenomena. Even more important, he emphasized the fundamentally creative and active character of human ensembles. The happy result of Mauss' work, unhappily little known in The New World, meets historians' objections that sociology abstracts the individual all too often from the total socio-historical context, and the psychologists' objections that the sociologist explains social phenomena by other social phenomena, never entering into the total context in which man alone

[11] *Ibid.*

has meaning. Georges Gurvitch, our great contemporary and a student of Mauss, carried the latter's work forward carefully and lycidly, but he is unfortunately little known in America.

All phases of the American scientific enterprise, both in the natural and the human sciences in the late nineteenth and early twentieth centuries were influenced more by Germany than by France, and Sociology is no exception, even though foreign influence has been only a minor factor. The founder of the first Department of Sociology in America (in the University of Chicago in 1892), Albion Woodbury Small (1854-1926),[12] had been a student in history at Leipzig during the early career of Wundt there. Perhaps there is some connection between American sociological empiricism and Wundt, as there is between American psychological empiricism and Wundt. At any rate, Germany was more important for American science than France, and a German—Weber, if not Wundt—helped American society interpret itself to itself.

Max Weber (1864-1920) was one of those who saw an unresolved problem in the foundations of the early science of sociology unnoticed by Durkheim: Durkheim's quantitative typology was inapplicable to total social phenomena, although Durkheim had hoped to achieve qualitatively distinct types within his sociology. Weber understood that only a valid concept of understanding (*Verstehen*) could serve as a method for constituting "ideal types" of social phenomena which would be qualitatively different.[13]

It is instructive to consider Weber's best known work in America, *The Protestant Ethic and the Spirit of Capitalism*, as a tract against Marx. The latter had maintained that the abolition of the alienating contradictions in the economic substructure of European societies would lead to new ideological positions. Weber maintained to the contrary that human options were the forces which shaped economies and societies, although his causal interpretation of human option is difficult to defend.

[12] Ernest Becker's *The Lost Science of Man*, New York, 1971, is an "outgrowth and extension" of the views expressed in his earlier *The Structure of Evil: An Essay on the Unification of the Science of Man*, New York, 1968. One of the essays in the book (which attempts to explain "the utter lack of a science of man in our time") is entitled "The Tragic Paradox of Albion Small and American Social Science." It masterfully demonstrates the paradigm-character of Small's career, and the failure of that career as the paradigm of the failure of American social science.

[13] The concept of *Verstehen* is the legacy of Dilthey. This is not the place to expand upon Dilthey's influence upon Weber—or indeed upon the whole history of ideas in Germany through Husserl to Jaspers and Schutz. It should simply be noted that Weber was a nominalist, not a phenomenologist, equating "comprehension" with "interpretation," and ignoring the "intuitive" character of *Verstehen* as understood by Dilthey.

American sociology developed in a somewhat different way than that of Germany, which is not surprising, given the differences in the societies in question. Weber (and Marx before him) had the remnants of a feudal society to deal with, while Americans were engaged in founding and guiding a new society upon its Road of Destiny. However, Weber's successor in German sociology, Karl Mannheim (1893-1947), maintained close connections with the English-speaking world, and made contributions to the sociology of knowledge which is becoming an important sector in the American sociological endeavor in our day.[14]

The foregoing paragraphs on early European sociology were intended not so much to indicate a direct influence of that work on American sociology as to indicate the kinds of problems in which the Europeans were involved. It is nevertheless true that Americans faced similar problems and their attempts to deal with them are not without resemblances to their European counterparts. The key figure in the implantation of sociology in America was the little-known American sociologist, Charles Horton Cooley (1864-1929), and it is not certain that any European Sociologist influenced his thinking, although his work shows striking analogies to that of Durkheim. In particular, his conviction of the irreducibility of social reality to any other reality and the irreducibility of collective behavior to individual behaviors mirrors both Durkheim's emphasis upon the "social" fact as fact and Durkheim's thesis that persons are drawn into organizations not because they have similar outlooks or activities but because their complementary *differences* constitute societies which manifest properties which are not idiosyncratic. Cooley identified "primary groups," on the one hand, which *are* founded upon likenesses among members, and "social organizations" which *are not* founded upon such likenesses. Cooley's American contemporaries were prone to see the "law of likeness" and the "consciousness of kind" among human beings as the unique source of group coherence.

In his endeavor, Cooley, with all his contemporaries, was attempting to find some "great law of human nature" which would account for the constitution of society. All the early sociologists were searching, like Comte, for *one* basic principle which would have a dual effect: that it have the aesthetic completeness of a truly scientific law and that it, at the same time, be generally applicable for major social problem-solving. In this search, let it be noted, the American penchant for problem-solving

[14] Since these paragraphs are written largely for the sake of suggestion we omit the important work of Simmel (1858-1918) and Vierkand (1867-1954).

was the more important factor. In the area of sociology, as in the area of psychological research, a Comtean positivist view guided early efforts. After the eighteenth century collapse of the religious world view, the nineteenth century had been faced with social movements and changes for which no model for control existed except the burgeoning model of nineteenth century scientism. The nineteenth century attempt to control everything through the miracle of science was translated into a social science which hoped, as the other sciences, to be truly scientific (thus positivist) and to construct a scientific model which could serve in the control of social change. In America, as one might have expected, the new sociologists followed the best scientific model of the day. At first there was the mechanist-evolutionary model; the mechanist model was imported directly from Europe, the evolutionary model was developed on these shores after the importation of Spencer. The attractiveness of Darwinian evolutionary theory as interpreted for the new science of sociology by Spencer, was its optimism and spirit of progress, a nourishing diet for the "melioristic" attitude of early American sociologists—the only attitude they held in common.

It is at this point, however, that the science of sociology as it was developed by its early practitioners shows its basic flaw. In any case, it would have been a difficult task to pursue the dual goal of sociology in a coherent way. Precise and well-controlled empirical social research could be reconciled only remotely with the humanitarian activities of the early American sociologists; although they argued much about "theory," the penchant of men like Ward, Ross, Giddings and Sumner was for the solution of social problems and not for precise sociological research.[15]

A new direction in American sociology was provided by the Chicagoan, William Isaac Thomas (1863-1947), to whom we are indebted for the still valuable concept of the self-fulfilling prophecy.[16] From 1896 on, he was in favor of full and concrete *evidence* in sociological research, as against the speculative argumentation prevalent in the sociology of those days. Consequently, his major contribution was the enormous field study of the Polish peasant.[17] In fact, however, Robert Faris pointed

[15] See Robert E.K. Faris, *Chicago Sociology: 1920-1932,* San Francisco, 1967, 3-7.

[16] Thomas's actual dictum was: "If men define situations as real, they are real in their consequences." It was his answer to the scholar's problem about whether "personal" statements could be of use to science (the basic problem being that of the reality-status of such personal statements)—a problem which still besets positivist endeavors in the human sciences today.

[17] William Isaac Thomas and Florian Znaniecki, *The Polish Peasant in Europe and America,* 5 vols., Chicago, 1918-1922.

out that the "Polish Peasant" study was two studies, one theoretical and one empirical; and, although this may seem odd, the massively impressive compilation of data was very little used in the theoretical elaboration. The effort at careful data-collecting, however, was to have far-reaching effects in American sociology, inspiring subsequent generations of American scholars.

This made the effort at attaining the dual goals of sociology all the more difficult. An instructive example of how the future of American sociology was to go is the history of the Chicago School itself. (The "Chicago School" is here taken as a paradigm of American sociology in the early days.) After Thomas's work had set an example of scholarship for his colleagues, the Department of Sociology was firmly committed to objective studies in the early 1920's. However, Thomas's influence was not negligible in the effort to solve social problems through social action. Social workers and others meeting Polish peasants at immigration offices were armed with an understanding of Thomas's scientifically solid hypotheses concerning the cycles of "disorganization" and "reorganization" which immigrants were encountering, rather than the traditional *a priori* prejudice that they were "biologically inferior."

The Thomas influence upon objective sociological studies culminates in the milestone work of Park and Burgess at Chicago.[18] As a result of the prestige of the Chicago School, of Park and Burgess themselves, and of Thomas to whom they were explicitly indebted, sociology as a science was firmly established and its methodological model well articulated. A new era in sociological research had begun, and sociology as a science was further advanced by careful and important studies issuing from the Chicago School. Park and Burgess collaborated on another, and still unsurpassed, study of the city.[19] There were further important studies of both urban and rural sociology.[20]

But the "Golden Age" of American sociology was relatively short-lived. Although the search for knowledge was a primary focus of the Chicago scholars, they were no less committed to applications of elaborated theories for human social welfare; their concern was that such applications be made on the basis of truly adequate empirical

[18] Robert Ezra Park and Ernest W. Burgess, *Introduction to the Science of Sociology*, Chicago, 1921.

[19] Robert Ezra Park, Ernest W. Burgess and Roderick D. McKenzie, *The City*, Chicago, 1925 (2nd ed., 1967).

[20] Pitirim A. Sorokin, C. Zimmermann, and C. Galpin, *A Sourcebook in Rural Sociology*, Chicago, 1930-1931; Robert and Helen Lynd, *The Middletown*, Chicago, 1929; *The Middletown in Transition*, Chicago 1936.

results and subsequent theory. Behind their commitment was the general spirit of the University of Chicago under the presidency of William Rainey Harper—first, its encouragement of open, modest, inductive, and anti-doctrinaire research in *all* fields; second, its general aspiration to pure science as interesting in itself; third, the conviction of the ultimate usefulness of all knowledge, provided one was open to following the lead of the investigation, without the pre-suppositions which might be programmed into it by funding institutions or interested governmental bureaucracies. This wise separation of "science" and "action" in the strictly *scientific* program was maintained with difficulty by the early scholars of the Chicago School, but maintained it was. Unfortunately, their successors were more interested, it seems, in the relatively easier task of data-collecting and statistics. Lynd and Sorokin were both to begin emitting "cries of alarm" rather early in the history of American sociology. Sorokin was to label much of the subsequent work in America as tainted with "testomania" and "quantophrenia" (as well as eleven other basic defects).[21] Self-critique of American sociology early became a fact of life; but, as we shall attempt to point out, critique has had little effect.

George Herbert Mead (1863-1931) is the first of the ambiguous figures in contemporary sociological theory. In America, it was pragmatism which overcame the dogmatic position of the mechanistic kind of social Darwinism—a feat it was unable to perform in the construction of mainline psychological theory; and it was George Herbert Mead, in the pragmatic tradition of Peirce, James, and Dewey, who reinstated man as the active subject in the processes of society. Although, as Winter admits, the background of Mead's work was behavioristic, still he challenged behaviorism's claim about the fundamental passivity of human consciousness and the submission of subjectivity to exclusively external conditionings. Winter feels that Mead's radical challenge made it impossible for behaviorism to gain control of the social sciences.[22] This, however, is doubtful. That is, it is doubtful that behaviorism has lost its hold on the social sciences[23] or that even if this were true, the credit should go to Mead.

[21] Pitirim A. Sorokin, *Fads and Foibles in Modern Sociology and Related Sciences,* Chicago, 1956; Robert Lynd, *Knowledge for What?* Princeton, N.J., 1939.

[22] Gibson Winter, *Elements for a Social Ethic: The Role of Ethics and Social Science in Public Policy,* New York, 1966.

[23] As late as 1964 William R. Catton Jr. reported that "the strategy of natural science in its broadest sense is now widely accepted as the effective strategy in Sociology." (See William R. Catton Jr., "The Development of Sociological Thought," Robert E.L. Faris, ed. *Handbook of Modern Sociology,* Chicago, 1964, 947.); yet *Sociology On Trial,* edited by Maurice Stein & Arthur Vidich, had appeared in 1963.

Surely, the pragmatic view is that man in an active problem-solver, that he shapes himself and his world, that the evolutionary process which Darwin disclosed finds man striving within that process. Winter is thus perspicacious to realize that, in the pragmatic view, Mead's project delves deeper than the mechanist view. The latter takes the evolutionary emergence of mind for granted, whereas Mead struggled precisely with the question of the emergence of the human mind within the evolutionary process as a result of an inter-subjectivity. Thus it is Mead who makes the link between pragmatism and the hopeful contemporary developments in sociology in which Winter figures so largely.

Mead's "social self" emerges in a social process; the "me" is myself as seen by the other. When I take the role of the other and see myself as he sees me, and when I assume the role of the "generalized other," he/it becomes internalized in me. Thus society becomes a part of the self, and the self, in turn, is socialized. The "social self," then, is "me," but the difficulty arises when "I," the operative self, become absorbed in "me"—the individual social result of collective society. Mead was not shallow; he began his thinking with the customary S-R model, which presupposes that stimuli have meaning, but moved to a model of "significant communication," in order to attempt to understand the very emergence of the meaning of the stimulus.

It seems, and has seemed for a long time, that "the self," in Mead's view, is only another instance of "social determinism"—a bit less gross than mechanism's utter determinacy, to be sure, but still "I" am absorbed in the "social self" ("me"); I become what society creates. "The social self becomes *the* self." [24] Mead, in thus reducing the self to the social self, unwittingly negated his own basic model, creating the basic paradox of his work, that "social selves emerge in social experience through communication, yet a social process of communication presupposes selves."[25]

On the positive side, Mead effectively broke social science away from mechanism, and moved it toward an organismic evolutionary model, following the path that Durkheim, Weber, and Cooley also trod; but Winter rightly concludes that Mead's social determinism triumphed, and attempts to reformulate Mead's theory so as to protect both creativity and determinism in social process.[26] It is unfortunate that for so long a time Mead's determinism *has* triumphed; this state of

[24] Winter, 23.
[25] *Ibid.*, 25-27.
[26] *Ibid.*, 31-32.

affairs was unnecessary, and has perhaps retarded fruitful thinking for thirty years. For example, studies on social interaction, in which human beings were considered as *real* actors in social change, were indeed undertaken in the spirit of *part* of Mead's theories, but they have received short shrift.[27] Another early attempt at the inclusion of human option within the ambit of social reality was the publication of a famous study of "social causation,"[28] but it has remained a neglected work for almost thirty years.[29]

It is no news that Mead's published output was quite small; and even his most influential book, *Mind, Self, and Society,* was published posthumously by his students from lecture notes and personal sketches. Only recently—as recently as the Spring of 1969!—has the creative part of Mead's thought come to light, again through the efforts of his former students. It will forever remain a mystery why Mead's literary executors did not see fit, in 1934, to publish the compensatory half of their master's work, to complete the diptych which would have made Winter's painful task of humanizing the human sciences in America unnecessary.

Mead's unresolved paradox, in fact, has had more serious consequences than one would have imagined. Had the balanced view which we *now* know was Mead's been available to sociologists during these many years, we might also have been spared the painful debate between the "functionalists" (thus Winter characterizes Parsons) and the "voluntarists" (read: Mills). The behaviorists, to be sure, would continue to count noses, but we could do without them in the real issue at stake in the debate: the ideological issue of the *control of society* under the canopy of *public policy*. At take in the passionately ideological debate between what Shils opprobriously calls "manipulative" and "alternative" orientation as against his own "consensual" orientation and

[27] Herbert Blumer, for example, wrote *Symbolic Interactionism: Perspective and Method,* Englewood Cliffs, New Jersey, 1969, in the spirit of Cooley for whom "man and society are born together." His reviewer and former student, Ralph Turner, (*Sociology and Social Research,* Jan. 1970) admits the force of Blumer's critique, yet maintains that such "symbolic interactionism" is still an enigma, even after thirty years. First, he says, others have not overlooked the manifest character of human action, but have deemphasized it "either because they concluded that interpretation-process was an epiphenomenon or they feel these processes cannot be directly observed, and that the closest attainable approximation to exact knowledge comes through the use of constructs known to be imperfect representations of reality." Further, the reviewer maintains, no concrete research has been done in Blumer's area.

[28] Robert M. MacIver, *Social Causation,* Boston, 1942.

[29] Heinz Eulau, "The Behavioral Movement in Political Science: A Personal Document," *Social Research,* 35 (1968), 1-29, says that the thirty-year neglect of *Social Causation* was probably due in part "to the fact that the technology of social science was insufficiently developed to confront the causal challenge."

procedure, and which Winter names rather "physical" (behaviorist) "voluntarist" (humanist) and "functional" (action-oriented) is precisely *what are—or are to be—our social values and goals.*

What has been happening, then, in American sociology during the past thirty years? A slow transformation of focus and method has gradually been at work. The effort of early sociologists—Durkheim, Cooley and even Park—to study the effects of social influences upon real human activity has been replaced by an effort to study an abstraction: "society." What were the contributing causes of this shift? First, there was the replacement of Comtean positivism with the positivism of the Vienna Circle in sociology as in all the human sciences in America. This introduction has had two effects: First, although science objectifies the world in order to study it, positivism in science must take the objectified reality for the lived reality of human being-in-the-world. This draws with it a second effect, the fiction of a "value-free" sociology. The fiction does not lie in the value-freeness of the objectified reality of social being (for positivism's "object" is indeed not subject to meaningful evaluation), but in the dissonance between this kind of objectification and the *de facto* evaluation of the social scientist that thus-and-so will *be* an object.[30]

A second contributing cause of the thirty-year detour of sociology was the adoption of an "organismic" model of social reality and its criterion of homeostasis, with the consequence that society could all too easily be anthropomorphized. Sociologists were led to attribute "needs" to the body politic as a whole, distinct from the "needs" of its members. The wise refusal to reject individual psychological factors as *sole* determinants of social behavior led to the opposite kind of reductionism, the refusal to accept *any* individual psychological factors. A society without people became the proper stable object of sociology, as a language without speakers became the proper stable object of Structural Linguistics.

Somehow a "movement" called "structural functionalism" (or simply "functionalism") has emerged. Structural functionalism as an "ism" is generally considered undefinable, even unidentifiable,[31] but functionalism as a method in sociology is identifiable and is called "functional analysis." It is characterized as follows: "One carries out functional analysis

[30] See T.S. Simey (Lord Simey of Toxteth), *Social Science and Social Purpose*, London, 1968; Gideon Sjoberg ed., *Ethics, Politics, and Social Research*, Cambridge, Mass., 1967; Alvin Gouldner, "Anti-Minotaur: The Myth of a Value-Free Sociology," *Sociology on Trial*, eds. Maurice Stein and Arthur Vidich, Englewood Cliffs, N.J., 1963.

[31] See Kingsley Davis, "The Myth of Functional Analysis as a Special Method in Sociology and Anthropology," *American Sociological Review*, 24 (1959), 757-772.

when, starting from the existence of a particular institution, one tries to find out what difference the institution makes to the other aspects of social structure."[32] Functional analysis, in some quarters, is considered *the* method of sociology, but even then it is not usually considered a theory; in other camps, those who accuse functionalism of having failed as theory because it shows the *interrelationships* between social institutions without explaining *why* these relationships are what they are, still acknowledge its validity as a *method*.[33]

The thirty-year dominance of the empiricism of functional analysis in sociology has reduced that science to nothing more than sociography. Homans criticizes the method in a purely practical way for not having explained anything; but from the point of view of philosophical critique, the problem seems more than "simply practical." The empirical description of social interrelationships (on individual or institutional levels) shows *how* institutions help maintain a society "in equilibrium," but implicitly, sociologists have assumed that "a society in equilibrium" is an optimum state of affairs, and wittingly or unwittingly have erected the equilibrium of the existing society into a *norm*.[34] If the sociography which issued from the sociological enterprise had alone been insisted upon, no great harm would have been done; indeed, the empirical description of societies and institutions would have met the first of the traditional demands of sociology as a science: the simple *desire to know*. However, the traditional demand of sociology was, and remains, dual: not simply the scientific desire for knowledge but also the pragmatic demand for *social action*. Once social equilibrium was, unfortunately, accepted as normative in the purely scientific part of the enterprise, a practical spin-off of social equilibrium was to become the normative goal of social action as well.

For this state of affairs, we have only to point to Talcott Parsons. Parsons argued that the study of social *change* must have some "reference point;" but this "reference point" eventually became the structure of the system, "the institutionalized patterns of normative culture."[35] The criterion of "institutionalization," then becomes "the patterns of normative culture;" but this leaves out of scientific consideration within the social system variant, deviant, or alternative structures or subcultures. These

[32] George C. Homans, "Bringing Men Back In," 811.

[33] *Ibid.*

[34] Talcott Parsons, "Some Considerations on the Theory of Social Change," *Rural Sociology*, 26 (1961), 219-239.

[35] See Walter Buckley, *Sociology and Modern Systems Theory*, 15.

must and do constitute part of social systems, but they can be dealt with only as deviant, both in sociological theory and in the effort at social action. Social coercion begins to loom upon the horizon.

Let us pause a moment here to review the basic skein of the discourse and to pich up dropped stitches, so as to be able to conclude these reflections in a coherent and hopefully telling manner. American sociology was early freed from "grand reform concepts for abolishing social evils," and "hereditarian bias and instinctivism." That is, "from excessive and unrewarding dependence regarding sociology on specialists in the field of Biology."[36] In sum, Park and Burgess had been successful in substantiating their principal theme: that human nature is a product of *social* living; the Spencerian evolutionary model was abandoned almost as soon as it was introduced in America, and the model of biological determinism was short-lived too. The attempt to explain how human societies cohere drew first mostly upon Tarde's doctrine of "imitation" (to the neglect of Tarde's own claim of the cyclic interplay of invention *and* imitation). Nevertheless, thanks to Cooley, and to Park's knowledge of Durkheim, early sociology in America recognized two bases for the organic unity of society: first, patterns of competitive relations which provide the basic material for the organization of complementary differences, second, similarities which provide the basic material for organized consensus. Of course, competition can (as Park thought it would) evolve through conflict, accommodation, and assimilation; but (as Park also understood) it can drive societies to lose out in the universal competition between societies because of what he called "their inner enemies—the defective, the dependents and the delinquents."[37]

Although the young science of sociology hoped to sever its link with the "do-gooders" (as Park called "meliorists"), Park himself, in his book with Burgess, wrote chapters on "social control" and "progress." The Comtean dream of achieving and controlling progress through scientific knowledge, including sociology, was thus still present in the adolescence of American sociology. But there is a fundamental difference between the program of dream-fulfillment in the early 'twenties and the contemporary version. The difference lies precisely in the matter of objectivity and positivism. Park was well aware that the "concept of progress is entangled with the values of the persons who use it," and thus liable to controversy and revision. Commenting upon a list of "indices

[36] Faris, *Chicago Sociology*, 14, 41.
[37] *Ibid.*, 37-50.

of progress" published by a statistician in 1913 (the list, incidentally, included such indices as "increase in population" and "racial homogeneity"), Park wrote, "Certainly, these indices are mere temporary measures of progress."[38] The concluding sentence of Park's book will help point up the contemporary impasse in Sociology: "From the point of view of social research the problem of progress is mainly one of getting devices that will measure all the different factors of progress and of estimating the relative value of different factors in the progress of the community." Park was thus willing to *estimate the relative value* of different factors in the progress of the community; however, the general present trend is antagonistic to such estimative evaluation. The only criterion for recognizing values *as values*, in mainstream Parsonian sociology, is the observable fact that some goals, norms, and standards, are *socially approved.*

Here, then, begins the predicament of the positivist: he is at a loss either to account for or to react to change in the socially accepted, observable, goals, norms, and standards of individuals, groups, or entire communities.[39] If the sociological positivist is true to his position, only two courses either of thought or action are open to him. Either he will consider novel and different values, norms, or standards (whether personal or collective) as *deviant,* and he will advocate "social control" as a legitimate tool to implement conservatism and conformity,[40] or he can refuse to take any cognitive stand with regard to novelty (if there is no observable "socially approved" value-model there can be no intelligibility), and simply wait until some kinds of "commonly accepted values" are again commonly accepted, so that he again has an "objective criterion" for recognizing values as values.

What would happen if the sociologist took one more small (and quite logical) step in making the definition of objectively identifiable "value" more precise? What if, instead of saying the value is merely "socially approved" norms and standards, he were to define value as these norms and standards *institutionalized*? In the area of analysis, he would isolate "his mechanisms of control and judge their relevance, not to the

[38] See Faris, 49.

[39] See Chapter VI, where the Structuralist, also the heir of Durkheim's "social fact," faces the same problem of accounting for change.

[40] Professional sociological journals are crammed nowadays with articles on social "deviations" of all kinds (with recipes on the use of behaviorist "learning theories" to cure the evils of society). Whatever one may think of "learning theories," their use indicates a shift in the focus of interest. The original program of the "humanitarians" in sociology was to cure the "evils" which society caused to men; today's programs, on the contrary, are largely designed to cure the "evils" which men do to society.

total system, but to that part represented by the dominant or legitimized structure that he has taken as a fixed point of reference."[41] In the area of social action, or social control as it is now called, he will favor any form of conformist education and of repression.

This is exactly the work of Talcott Parsons. Although Parsons has had enormous influence upon sociology, his work has been consistently criticized for more than twenty years—without apparent effect. Buckley apologized for undertaking "the tedious job of adding yet another to the many critiques of the Parsonian framework."[42]

Buckley himself represents a recent movement in sociology which attempts to come to terms with the extreme inadequacy of the dominant Parsonian model, a movement allied with a theory of "general systems." Buckley's work, however, is based upon a much more far-reaching critique than a simple refutation of Parsons; he attacks all the historically accepted models in social theory—equilibrium, consensus, and functional theories as well as the more general organismic and mechanical models. He seeks, on the basis of information theory and cybernetics, to counteract the useless empirico-analytic procedures of classical sociology.

As a true social *scientist* Buckley is concerned primarily with understanding the complex system which is our modern society; he includes consciously within the system (since they *are* in the system) all varieties of social novelty, admitting that some variances may be pathogenic, but not by the mere fact that they are variances. Buckley is fully aware that variations from the "common norm" will have to be appraised and evaluated, but does not seem to think that is the task of the sociologist, and he is not a social actionist. (However, by admitting variances *within* his model, Buckley's attitude of tolerance might be considered "humanitarian.") Thus, he attempts to fulfill at least half of the dual task of American sociology—the actualization of the pure desire to know. But, within the investigative procedures which he undertakes, there is an element hithertofore unnoticed which comes to light, with enormous consequences:

> To correct this onesidedness (of Parsonism) and be consistent with the realities of a social system, it would seem necessary to take the structured strains and structured deviance as an integral part of the system and a fixed point of reference and ask what mechanisms may be isolated out that tend to maintain *these* structures.

[41] Buckley, 30.

[42] In according such scant attention to one of the accepted Father Figures of American Sociology, we are aware of having been preceded in this negligence by Georges Gurvitch, who makes no mention of Parsons in his long chapter on the history of sociology, and remarks elsewhere in his book simply that Parsons' efforts were a "complete failure."

The answer would lead straight to the crucial problems of power, ideology, and propaganda, vested interest, and the like.[43]

Recognition of the imperious need to evaluate variances, norms, and values is spreading, since in the ideological battle raging within the confines of the social sciences there are issues at stake which extend far beyond the limits of these sciences: precisely the control and regulation of society itself. Gibson Winter has pointed out the tremendous impact of the social sciences in the formation of public policy at all levels—economic and educational policies, international politics, and even global military strategy. The influential character of the social sciences in our culture, from "urban renewal" to lunar debarcations, can hardly be contested. The social scientist, together with the psychologist, has become the sage of the twentieth century; his counsel is sought in the seat of all governments, he participates actively in the formation of public policy regarding everything from the neighborhood street and its inhabitants to the stance of the body politic regarding its real or supposed enemies and friends. This assessment is not fanciful; it is too mild to fit even the few facts visible at the top of the iceberg.[44]

What models do the social scientists offer as maps for the solution of the problems which face us—or even for the designation of something as a "problem?" Irene Taviss has outlined some of the challenges for the social sciences (including economics, political science and sociology) which are inherent in our increasingly technological society.[45] She clearly recognizes the basic challenges for social sciences today, but sees just as clearly that there is no master plan for solving economic, social, and political problems, or even for identifying the problems. We proceed, she hypothesizes, from a vague conviction about "inbred notions of the good." From this, we derive, somehow, a selected number of socio-economico-politico-psychological "desirables." These "desirables" are then translated into a set of problems (which presumably deal with *real*

[43] Buckley, 30. Parsons, he is saying, can consider (in terms of his model) "only mechanisms of defense, adjustment, and deviance control, all aimed at adaptation of the *actor* to a given dominant structure, with no consideration given to the historically obvious mechanisms that adapt or change the system structure to accommodate the actor and maintain the total system." European observers, in fact, notice a curious phenomenon in American sociology: that there is much consideration of "initiative," but no serious studies of *power* and *legitimacy*. Some American sociologists are prepared to admit that there are some areas of American social life which *cannot* be studied.

[44] See the report on the scandalous "Life and Death of Project Camelot," in Irving Louis Horowitz, *Professing Sociology: Studies in the Life Cycle of Social Science*, Chicago, 1968, 287-304.

[45] Irene Taviss, "The Technological Society: Some Challenges for Social Science," *Social Research*, 35 (1968), 521-539.

problems, to the extent that these can be re-translated into the terms of the "science" in question). Somehow, with these problematic schemata, we attempt to answer the policy-maker's demand for help in "planning." The social scientist, she admits, cannot give a satisfactory operational meaning (after all, the only possible "scientific" meaning there is) for the desirables of our society, nor can he answer the question about their intrinsic worth (except in terms of his own inbred notion of the good).

Current literature in the human sciences (here Miss Taviss and Gibson Winter agree remarkably well) is discussing the "relative failure" of the current models, both the "functional" models which are based on the idea that values are the shared definition of the desirables which make societies hold together, and "conflict" models, which are based on the idea that values are the emanation of the interest of ruling classes or groups. This is not surprising; Edgar Morin has incisively shown that *all* models in the social sciences are broken, that "anthropolitics" is reduced to bits and pieces.[46] Current literature, according especially to Miss Taviss, reflects this "inconclusive and confused" behavioral type in scientific outlook when faced with what Goldstein would have called a "situation of catastrophe." Some social thinkers consider the "values" in question epiphenomenal or scientifically unknowable. This stance *might* be theoretically defensible, and interesting enough in academic debate, were it not for the presence of these same social thinkers in the seat of arbitration over public policy. Others, on the other hand, can save scientific methodology from shame by postulating that behavior is an expression of values. Still others see the theoretical answer as somewhere between these two extremes. The general current conclusion is a general agreement that "values tend to shape goals, and goals, in turn, influence behavior." Miss Taviss here puts her finger upon the heart of the real problem: there is no current successful attempt to explicite "scientifically" the linking mechanisms of these three factors, no valuable studies of value-formation, and no question about the *nature* of "values." We seem to be back at the starting point of an impressive "scientific" study of man-in-society: "inbred notions of the good." It is not difficult to see that the *ideological* use of such doubtful "scientific facts" as can be scratched up by the social scientist can have horrendous effects upon the society in which they are supposedly apprehended and for which they are used. Perhaps this pill is more easily swallowed if proferred in Winter's more elegant assessment of the dilemma of the

[46] Edgar Morin, *Introduction à une politique de l'homme*, Paris, 1969.

current positivism in social science: "it confuses its own abstractions and formulations with social process (the error of reification of theory) but in turn it sets its formulations of the real law of the process over against the interests of society."[47]

Winter asks what differentiates "science" from "ideology," since "science" is never pure, always has interests to protect, if only those of the scientists involved in it. The human sciences, above all, are the ones the most closely engaged within the phenomena they are supposedly "objectively" appraising. This is perhaps the reason why power and letigimacy are not analyzed by sociologists.[48] Rather than attack the human sciences for being ideological, however, he asks positively about what is really scientific in the explorations of the human social enterprise, and remarks astutely:

> Such questions precede any evaluation of the various models of the social process which are proposed by different groups within the human sciences. The prior question of the meanings of a scientific attitude does not resolve the question of adequate models, but defines the limits of relevance of any theoretical reflection in apprehending the nature of man's practical activity.[49]

His effort, then, is to find the grounds of science in the living experience of the human person, including the human scientist; and he uses a phenomenological method to distinguish science from ideology. He retraces, in a fine way, the steps which led Husserl from his critique of the commonsense world to his statement of the crisis of Western Science. With the help of Schutz, he reestablishes the balance between human creativeness and social conditioning, passes on to answering the question of the possibility of evaluation of scientific models, thus laying the basis for a social ethics.

Winter's book, which he considers a "first meager step," is truly a milestone in the field of social ethics, and a masterful commentary upon the positivist model of society and the difficulties it poses for our common life. The reader can do no better than retrace with him the steps which he makes there.

[47] Winter, 12.

[48] In contrast with thousands of studies of social deviance, those on power and legitimacy are easily counted. Arnold Rose, in his *The Power Structure: Political Process in American Society,* New York, 1967, cites only two other works on this topic: Floyd Hunter, *Community Power Structure: A Study of Decision Makers,* Durham N.C., 1953; and C. Wright Mills, *The Power Elite,* New York, 1956. The others, he says, are done rather by social historians and political scientists. The writer finds only one more example, Gideon Sjoberg (ed), *Ethics, Politics, and Social Research,* Cambridge, Mass., 1967.

[49] Winter, 52.

For our part, the task is rather to discuss the very notion which Winter described so well: the foundation of science in the world of everyday experience.

Even Buckley, be it noted, though promising an "operationally definable, objective, non-anthropomorphic study of purposiveness, goal-seeking system behavior," and so forth, cannot successfully meet the objections of the philosopher Richard Taylor, that this "goal-seeking," from the observer's point of view, might apply to anything in motion, even a rolling stone. Thus, the general systems theoretician, for all his helpfulness in partially overcoming the positivist dilemma, is himself using the ordinary, human, pre-philosophical and pre-scientific knowledge we all have of what it means to "strive" in the world.[50]

"Objectivist" social scientists have simply overlooked what they are, in fact, drawing upon such pre-scientific and pre-philosophical commonsense knowledge not only in the interpretation of "data" but in the very enterprise of data-gathering. The scientific models of society, social order, and social process have now reached the limit of their effectiveness for dealing with the problems which they face and which they helped to create. They present, even to the social scientist, an "inconclusive and confused image" of society and of the place of sociology within it. It is now recognized more and more widely that it is time to get "back to the things themselves."[51]

It is here that once again phenomenology is coming to the rescue of the human sciences, including sociology. Let us simply repeat that it was the merit of Husserl, through a lifetime of patient and unfinished work, to have rediscovered the subject in his subjectivity (long before Homans suggested to sociologists that it was time to be "getting men back in"); it was Husserl who realized that the everyday world in which we live is not the "reality" of the sciences, that we live real lives in parallel with scientific theories about reality, and that our real and everyday experiences provide the foundation of possibility for constructing scientific theories about social experience. No one lives in the "world

[50] Buckley, 35; 69-70. See also Richard Taylor, "Purposeful and Non-Purposeful Behavior: A Rejoinder, "*Philosophy of Science*," 17 (1950) 327-332, for references to the whole controversy. (Taylor's argument was with Rosenblueth, Wiener and Bigelow.)

[51] See Norman MacKenzie (ed.) *A Guide to the Social Sciences*, New York, 1966; C. Wright Mills, *The Sociological Imagination*, New York, 1959; Arnold M. Rose, "Varieties of Sociological Imagination," *American Sociological Review*, 34 no. 5, October 1969; Robert A. Nisbet, "Sociology as an Art Form," *Pacific Sciological Review*, Fall 1962; Maurice R. Stein, "The Poetic Metaphors of Sociology," *Sociology on Trial*, Maurice Stein and Arthur Vidich eds., Englewood Cliffs N.J. 1965, 173-182; J.B. Mays, "The Poetry of Sociology", *University of Liverpool Inaugural Lecture Series*, Liverpool, 1968.

of science," and Husserl's great achievement was to have reminded us of this fact.[52] It was in the context of Husserl's phenomenological thinking that Winter, following Schutz to reinterpret Mead, was able to present the elements of a social ethic which respects both a scientific perspective and the creativity of everydayness.

Alfred Schutz is the key figure in the renewal of sociology. His life and career has been briefly and touchingly sketched by Herman Van Breda in his preface to the first volume of the *Collected Papers of Alfred Schutz,* tracing the steps of his scholar friend through his early studies in law and the social sciences (under Ludwig von Mises, among others), when the controversy over the definition and distinction of *Naturwissenschaften* and *Geisteswissenschaften,* now happily and fruitfully settled, was under discussion, and the influence of Dilthey, Rickert and Weber upon Schutz's early thinking. After his contact with Husserl's phenomenology, Van Breda reports, Schutz's aim in life became phenomenological: "To found philosophically the understanding of meaning as actually employed and understood in the human sciences is to find its source in the life of consciousness itself."[53]

Husserl's theory of intentionality, of intersubjectivity, and of the *Lebenswelt* were the guideposts of Schutz's thought, as they have become those of Winter in his conciliating efforts. Maurice Natanson, editor of the first volume of Schutz's papers, remarks in his introduction that "the philosophy of Alfred Schutz articulates a single intuition, the discovery in full depth of the presuppositions, structure, and signification of the common-sense world . . . the *fil conducteur* of his intellectual life was a concern for the meaningful structure of the world of everyday life . . ."[54] And again:

> To see this world in its massive complexity, to outline and explore its essential features, and to trace out its manifold relationships were the composite parts of this central task, the realization of a philosophy of mundane reality, or, in more formal language, of a phenomenology of the natural attitude. The understanding of the paramount reality of commonsense life is the clue to the understanding of the work of Alfred Schutz.[55]

Under the heading of "the commonsense world," Natanson explains, Schutz found the *a priori* conditions of its possibility in one's biographical

[52] See Robert Welsh Jordan, "Husserl's Phenomenology as an 'Historical' Science," *Social Research,* 35 (1968), 245-259, for The suggestion that even Heidegger's treatment of historicity is *directly* dependent upon Husserl's thought.

[53] Alfred Schutz, *Collected Papers,* Vol. I, *The Problem of Social Reality,* The Hague, 1962, p. viii.

[54] *Ibid.,* p. xxv.

[55] *Ibid.*

situation, in one's available stock of knowledge, and the coordinates of the social matrix of "my" situation based on the place which my body occupies in "my actual Here." The intersubjective element in my situation implies a "there" as well as a "here," and draws with it the question of the "alter ego" and the whole sequence of alter egos which is implied in the historicity of my own ego. For present purposes, perhaps the most important facet of Schutz's thought is his concern with the "understanding of social action as the meaning which the actor bestows upon his action, i.e., the meaning his action has for him."[56] Schutz takes his departure from Max Weber's postulate of the subjective interpretation of meaning, but enlarges and deepens it through his use of Dilthey's idea of *Verstehen,* understood as interpretive understanding, as the key to the understanding of human life and activity. *Verstehen* is the epistemological access to the *Lebenswelt* which Husserl suggested as the name for "the Life-world that encompasses the rich totality of commonsense experience lived through by the individual in his concrete existence."[57]

Even from this all too summary and incomplete account of Schutz's phenomenological enterprise, one can already see not only the corrective for the narrow incompleteness of the positivist influence upon the human sciences, but also the dawning hope for the reinstatement of the human subject in the sciences which deal with him, where a basic reflection upon the *foundations* of both philosophy and all positive sciences can permit them to pursue their several tasks honestly and fruitfully—and even, perhaps, cooperatively.

We recognize that this phenomenological effort represents a *new* kind of sociology, taking its point of departure not from the "social world" as objectified, but from that social world as lived by human beings. It has been inspired by Alfred Schutz in the past few years, although if American sociology had taken seriously the "symbolic interactionism" of Blumer (drawing his inspiration from Cooley, Mead, and Dewey), it would probably have arrived at a phenomenological stance of its own. We have spoken at length earlier in this book of the solidity of a phenomenologically based "third force" in psychology. This same kind of "third force" is now gathering in the field of sociology as well, and we can probably soon expect a "Journal for Humanistic Sociology"

[56] *Ibid.,* xxxv.

[57] *Ibid.* Natanson, who has been very helpful here, explicitates three kinds of relationships between *Verstehen* and the *Lebenswelt.* See also his "A Study in Philosophy and the Social Sciences," *Philosophy of the Social Sciences: A Reader,* Maurice Natanson (ed.), New York, 1963, and "Alfred Schutz on Social Reality and Social Science," *Social Research,* 35 (1968), 216-244.

although tentative efforts in that direction presently find hospitality in the "Newsletter" of the Association for Humanistic *Psychology*.[58]

How then can we conclude these reflections? Perhaps very briefly by simply pointing out that the "Model of Man" profered by mainline American sociology is deeply contradictory. Its researchers and theoreticians have followed through with Durkheim's conclusion that man is a product of society, neglecting Cooley's insight that "man and society are born together," but at the same time have hoped and planned social action (either to change man's social condition or to coerce him into conformity). Either the planners are another brand of human beings, capable of initiating action and change, or, if they are like the rest of men, they and their procedures are self-contradictory.

However, with all its failures, mainline sociology has paved the way for a new start.

By demonstrating our conditionings, sociologists themselves have shown that:

> Unlike . . . puppets, we have the possibility of stopping in our movements, looking up, and perceiving the machinery by which we have been moved. In this act lies the first step towards freedom. And, in this same act, we find the conclusive justification of Sociology as a humanistic discipline.[59]

[58] See John F. Glass, "Toward a Humanistic Sociology," *Newsletter of the Association for Humanistic Psychology,* 6 no. 4, April 1970. We must not, however, neglect to point out that this is the orientation of *Social Research* published at the New School for Social Research and edited by Peter Berger, and that it was the New School which introduced Schutz to America. He taught there from 1939 until his death in 1959. Although the publication of his major works in English is quite recent, his colleagues and students have already begun to make his thought known in America. The most prominent are Aron Gurwitsch, Peter Berger and Maurice Natanson.

[59] Peter Berger, *Invitation to Sociology: A Humanistic Perspective,* Garden City, New York, 1963, 176.

The reader has undoubtedly formed a judgment concerning the inadequacies of this chapter. The writer is not unaware of them, but would like to think they stem from an excessive allusiveness. Part of the project in this essay was to show respect for the truly scientific character of the human sciences. Regarding a sociological Model of Man, however, a certain uneasiness was unavoidable; but the author is not a sociologist, and restraint was imposed. Since this text was written, however, Robert W. Friedrichs' *A Sociology of Sociology* appeared. Using Thomas Kuhn's *The Structure of Scientific Revolutions* as a framework, Friedrichs pitilessly unmasked the sham value-neutrality of the reigning "priestly" paradigm in American sociology, and indicated the ethical dimensions of the clandestine options of social scientists. The writer's review of this book will appear in the international journal, *The Human Context,* in 1971.

More recently, David Bakan, in an unpublished paper entitled "The Social Context of Psychology," demonstrated the almost universal failure of the "applied" sciences, and traced this failure to the complacent "two-step" self-mystification of scientists.

If these judgments stand up, perhaps the respect we have all hitherto shown for "Science" will have to be reevaluated, not only in epistemological but also in moral contexts. Even so, the thesis of this essay that there are unconscious or unavowed options at the basis of all sciences and philosophies can stand.

SELECTED ADDITIONAL READINGS

FOR CHAPTER IV

Bierstedt, Robert, "The Common Sense World of Alfred Schutz," *Social Research,* 1963 (30), 116-121.

Bourdieu, Pierre, "Structuralism and Theory of Sociological Knowledge," *Social Research,* 1968 (35), 681-706.

Buckley, Walter, ed., *Modern System Research for The Behavioral Scientist: A Source Book,* Chicago, 1968.

Cooley, Charles Horton, *Sociological Theory and Social Research,* New York, 1930.

De Laguna, Grace A., "The *Lebenswelt* and The Cultural World," *The Journal of Philosophy,* 1960 (57), 777-791.

Durkheim, Emile, *Suicide: A Study in Sociology,* trsl. John A. Spaulding and George Simpson, Glencoe, Ill., 1951.

—, *The Elementary Forms of The Religious Life: A Study in Religious Sociology,* trsl. Joseph Ward Swain, 2nd ed., New York, 1965.

Emmet, Dorothy, and Alasdair MacIntyre, eds., *Sociological and Philosophical Analysis,* New York, 1970.

Etzioni, Amitai, *The Active Society: A Theory of Societal and Political Processes,* New York, 1968.

Friedrichs, Robert W., *A Sociology of Sociology,* New York, 1970.

Glaser, Barney G., and Anselm Strauss, *The Discovery of Grounded Theory: Strategies for Qualitative Research,* Chicago, 1967.

Gouldner, Alvin W. *The Coming Crisis of Western Sociology,* New York, 1970.

Gurwitsch, Aron, "The Common Sense World as Social Reality: A Discourse on Alfred Schutz," *Social Research,* 1961 (28), 71-93.

Hatley, H.G., "Limitations of Systems Analysis," *The Phi Delta Kappan,* 1969 (50), 515-519.

Honigsheim, Paul, *On Max Weber,* trsl. Joan Rytinia, New York, 1968.

Horowitz, Irving Louis, ed. *The New Sociology: Essays in Social Science and Social Theory in Honor of C. Wright Mills,* New York, 1964.

—, *Professing Sociology: Studies in The Life Cycle of Social Science,* Chicago, 1968.

Lazarsfeld, Paul, William H. Sewell, and Harold L. Wilensky, eds., *The Uses of Sociology,* New York, 1967.

Lundberg, George A., *Foundations of Sociology,* New York, 1939.

Manis, Jerome G., and Bernard N. Meltzer, eds., *Symbolic Interaction: A Reader in Social Psychology,* Boston, 1967.

Maslow, Abraham H. "Some Fundamental Questions That Face The Normative Social Psychologist," *Journal of Humanistic Psychology,* 1968 (8), 143-154.

Mead, George Herbert, *Mind, Self, and Society: From the Standpoint of a Social Behaviorist,* Chicago, 1934.

Means, Richard L., *The Ethical Imperative: The Crisis in American Values,* Garden City, N.Y. 1969.

Merton, Robert K., Leonard Broom, and Leonard S. Cottrell, Jr., eds. *Sociology Today: Problems and Prospects,* New York, 1959.

Mills, Charles Wright, ed., *Images of Man: The Classic Tradition in Sociological Thinking,* New York, 1960.

Natanson, Maurice, *Literature, Philosophy, and The Social Sciences: Essays in Phenomenology,* The Hague, 1962.

—, "Alfred Schutz on Social Reality and Social Science," *Social Research,* 1968 (35), 217-244.

—, "Phenomenology and Typification: A Study in The Philosophy of Alfred Schutz," *Social Research,* 1970 (37), 1-22.

Nisbet, Robert A., *The Social Bond: Introduction to a Study of Society,* New York, 1970.

Parsons, Talcott, *The Structure of Social Action,* 2nd ed., Glencoe, Ill., 1949.

—, and Edward A. Shils, eds., *Toward A General Theory of Action,* New York, 1962.

Psathas, George, "Ethnomethods and Phenomenology," *Social Research,* 1968 (35), 500-520.

Rose, Arnold, ed., *Human Behavior and Social Processes,* Boston, 1962.

Schutz, Alfred, *Collected Papers,* Vol. I: *The Problem of Social Reality,* ed. Maurice Natanson, The Hague, 1962.

—, *Collected Papers,* Vol. II: *Studies in Social Theory,* ed. Arvid Broderson, The Hague, 1964.

—, *Collected Papers,* Vol. III: *Studies in Phenomenological Philosophy,* ed. Ilse Schutz, The Hague, 1966.

—, *The Phenomenology of The Social World,* trsl. George Walsh and Frederick Lehnert, Evanston, Ill., 1967.

Shibutani, Tamotsu, ed., *Human Nature and Collective Behavior: Papers in Honor of Herbert Blumer,* Englewood Cliffs, N.J., 1970.

Sjoberg, Gideon, ed., *Ethics, Politics, and Social Research,* Cambridge, Mass., 1967.

—, and Robert Nett, *A Methodology for Social Research,* New York, 1968.

Stein, Maurice, *The Eclipse of Community: An Interpretation of American Studies,* Princeton, N.J. 1960.

Stinchcombe, Arthur L., *Constructing Social Theories,* New York, 1968.

Tiryakian, Edward A., "Existential Phenomenology and The Sociological Tradition," *The American Sociological Review,* 1965 (30), 674-688.

Tönnies, Ferdinand, *Community and Society,* trsl., Charles P. Loomis, 2nd ed., New York, 1963.

von Bertalanffy, Ludwig, *General Systems Theory: Foundations, Development, Applications,* New York, 1968.

Weber, Max, *The Methodology of The Social Sciences,* trsl. and ed. Edward A. Shils and Henry A. Finch, Glencoe, Ill. 1949.

—, *The Protestant Ethic and The Spirit of Capitalism,* trsl. Talcott Parsons, New York, 1958.

Wolf, Kurt, H., ed. *Emile Durkheim: Essays on Sociology and Philosophy, With Appraisals of His Life and Thought,* New York, 1964.

Zaner, Richard M., "Theory of Intersubjectivity: Alfred Schutz," *Social Research,* 1961 (28), 71-93.

CHAPTER V

THE MARXIST MODEL

THE DREAM OF THE "NEW MAN" AND A RUDE AWAKENING

Who is, then, the total man? Neither physical nor physiological nor psychological, nor historical, nor exclusively and unilaterally economic or social. He is all that, and more than the sum of these elements or aspects. He is their unity, their totality, their becoming. Man is defined by his knowledge, his sciences, by that which the sciences discover; but the sciences themselves are determined by man acting and thinking. While the old scientism is content either to privilege abusively one or other science and to conceive everything physically, for example, or mathematically or biologically, or to envisage a sum of the results acquired by the various sciences, dialectical materialism puts man in the center of its preoccupations...

The most widely propagated error (voluntary or not) consists in confusing historical materialism (dialectic) and a vulgar materialism (mechanism). The latter reduces the nature of matter brutally defined by its mechanical properties (volume, density, elasticity, etc.). It reduces beings of nature to mechanical combinations of these elementary properties (combinations of particles, of corpuscles; atomism). It reduces thought to a secretion, and consciousness to an epiphenomenon of physiological or physio-chemical processes. It reduces that which is human to elementary organic needs... This reduction from the complex to the simple, from the superior to the inferior, ends up with an extremely impoverished conception of the world and of man.

It must be remarked that this theory, long surpassed by physics and natural science, still is current in certain human sciences (Watsonian Behaviorism — the Social Darwinism of Spencer....)

These lines, which (with few changes) could have served as a statement of the thesis of the present volume, are found under the pen of the indefatigable French Marxist pamphleteer, Henri Lefebvre.[1] The statement is all the more surprising when one considers that absolutely no echo of Lefebvre's statement of the essence of dialectical materialism can be found in the longwinded discourses of the masters of the Kremlin, the "Concordat of Bratislava" in which the Czechoslovak leaders agreed to the Russion invasion of their country in 1968, or the Little Red Book

[1] Henri Lefebvre, *Le Marxisme*, Paris 1969, 112, 107-108.

of Mao Tse Tung—not to mention any other of the contemporary manifestos of the various official Communist Parties.

Part of the surprise is diminished by the simple realization that Marx's contribution to Marxism has been revised and his battle against ideologies has itself been turned into an ideology, and that, as Trotsky predicted, once the interests of the *proletariat* (the social class which Marx claimed has in itself the only possibility of social, economic, and personal change) become invested in a *political party*, that party is open to a takeover by a dictatorship which soon re-establishes the original situation of the oppressed proletariat. When Trotsky was pickaxed to death in a Mexican prison, his prophecy was already coming true; and Marxist ideologists have been struggling ever since to come to terms with the Leninist-Stalinist revisions of the original program of Marx, or to proclaim those revisions heretical (as do the Chinese). We shall have to return to these themes later; these few lines are an attempt to give a perspective to the paragraphs which follow.

Apart from the dynamics of national and international power politics, what accounts for the wide divergence of Marxist theory—whether philosophical or ideological? Basically, it seems, the starting point of the confusion is found in the real or apparent contradiction in the work of Marx himself—either in terms of comparing the *Manuscripts of 1844,* the work of the young Marx (that attractive young man, someone remarked, who perhaps never existed) and the work of his mature years, the *Grundrisse der Politischen Oekonomie,* of 1857-1858, and *Das Kapital* (1867), or in terms of the incoherencies present already in the early works. In this effort, the young Marx is seen, though ambiguously, as a humanist-anthropologist, the older Marx as a historico-revolutionary economist. The basic problem is whether the theory of alienation-disalienation proposed, however tenuously, by the young Marx is inevitable in *any industrial society*, and thus whether the classic "stages" of society later formulated must be fulfilled in *every* case.

This question will, perhaps, be clearer if we look more closely at the evolution of Marx's thought and the reasons for the contemporary controversies about its meaning. In his *Manuscripts of 1844* Marx discovered the historical roots of work as alienation, the exploitation of man by man, and sketched out its origins, its deployment, and the conditions of its disappearance. In the *German Ideology* (with Engels) of 1846, the root of alienated work is made precise: it is the division between work and the production of merchandise. In his *Capital* the fetish-character of economic categories is analyzed through investigation of relationships between various kinds of "merchandise" (of which *work* is one) with the result that private property and market competition are seen as deployed across the board, isolating and alienating "producers" and "owners" from one another even before the appearance of capitalism, from the very beginning of the production of saleable merchandise even in the most primitive societies.

Everyone seems to agree that the evolution of the concept of work as alienation in Marx' writings begins with an *anthropological* concern (derived from Hegel as revised by Feuerbach), and ends with a *historical* concept of alienation, the *Manuscripts of 1844* serving as a transition. Contemporary controversy over the fact of this transition and its significance began with the "re-readings" of Marx in various contexts (neo-Kantian, Bergsonian, neo-Hegelian, phenomenological, existentialist, theological, structuralist), with the common aim of reinterpreting Marx in the light of the works of his youth.

There are, at present, three viable positions; First there are those who find no basic difference between the *Manuscripts of 1844* and the *Capital*, and who find, rather that the fundamental theses of this position are held, strangely enough, by some officials of the Communist Party (e.g. Togliatti and Garaudy [2]) as well as by "ferociously" anti-communist Socialists like Erich Fromm, and by Catholic theologians such as Bigo and Calvez. A second group attempts to reevaluate the *Capital* in the light of the earlier works (thought to be more true to Marx's real thought on the alienation of labor), giving an anthropological, purely philosophical or even ethical interpretation of the concept of alienation. This group, which has had the loudest voice in the discussion up to now, includes, among others, people like Karl Löwith, Herbert Marcuse, and Kostas Axelos.[3] The third group, represented by the official ideologues of the Communist Parties of the 1940's and 1950's and by Auguste Cornu, the biographer of Marx and Engels [4] maintains that the *Manuscripts of 1844* mark only a transitional stage in Marx's thought. On this reading, Marx

[2] Roger Garaudy, one of the principal sources of the ideas of this chapter, was a Barthian Christian in his youth. He was a member of the French Communist Party for thirty-six years, and—until his exclusion in February 1970—twenty-four years a member of its Central Committee. It is not clear whether his general "humanist" tendency or his opposition to the Russian occupation of Czechoslovakia contributed more to his excommunication. (Garaudy was expelled from the French Communist Party in May, 1970.)

[3] Ernest Mandel, *La formation de la pensée économique de Karl Marx*, Paris, 1967, 165. See also 151-180.

[4] Cornu is cited by Mandel, 170. Cornu's monumental work, *Karl Marx und Friedrich Engels: Leben und Werk*, in three volumes, received highest praise from Garaudy as "an indispensable working instrument," and the only such work motivated exclusively by scientific (i.e. non-ideological) preoccupations. (See Garaudy, *Karl Marx*, 24).

already understood the basic contradictions of bourgeoise society, but expressed them in the humanistic language of Hegel and Feuerbach. Later, they say, Marx freed himself from the humanistic thinking of his youth and went on to formulate his theory of value and plus-value independently of it, with the consequence that any philosophical (anthropological or other) explanation of history and man is impossible. Mandel refutes *all* these positions, finding that Marx's original intention was a critique of the entirety of bourgeoise society, that his first step was the formulation of some general laws of social evolution, one of which consists in saying that production-relationships constitute "the anatomical system" of society, that the adequate formulation of this law required the assimilation of *all* the human sciences, that, consequently, Marx could only understand economics by understanding that science in the context of all the sciences, with the further consequence that it was Marx alone who could bridge the gap between the *history* of economics and the *analysis* of economics. To put this in structuralist terms, it was Marx who successfully unified the synchronic study of society (of which economy is the basis) with the diachronic study of the same phenomenon. Mandel remarks that "it is impossible to *dissociate* in Marx the sociologist from the revolutionary, the historian from the economist." He concludes that Marx "could not have been efficaciously, that is, scientifically, sociologist, historian, and above all revolutionary, unless he had been an economist..." [5] Thus, in a sense, Mandel's assessment of Marx greatly resembles Richardson's assessment of Heidegger; as Heidegger I could not be without Heidegger II, so there could be no Marx I (of 1844) without Marx II (of 1857).[6]

Before proceeding with a more detailed study of the passage between the "two Marxes" it may be helpful to interpolate a few remarks concerning the difference between "ideology" and "philosophy." Although the distinction will have to be drawn more finely before we finish, it

[5] Mandel, 203.

[6] There is, so far, no appropriate way of dealing with Marxism from a phenomenological point of view, unless the wide-ranging hermeneutic of Gadamer can be called in for help. These pages, then, are only preliminary to a phenomenological critique. They attempt a first approximation at uncovering the intentionalities of Marx himself, in contrast to socialist ideologues and neo-Marxists. A further step would entail a thorough analysis of the history of the "Frankfort School" and the contemporary *"Methodenstreit"* there between Horkheimer and Adorno on the one hand, and Habermas on the other. One of the key points of discussion will have to be the work of the disappointed revolutionary, Herbert Marcuse (who keeps alive the original rationalization that "facts cannot contradict the truth" of the inexorable laws of history) and Habermas' contention that in the last analysis it is "interest" which guides even the most abstruse theoretical reflections. The possibility of general historical laws and a consideration of Popper's lifelong battle against "historicism" would then take second place.

might be enough to remark, in the words of Bottigelli, who "presented" the French translation of Marx's *Manuscripts of 1844,* that "Marx resolved concretely the problem of the identification of the subject and the object which Hegel had resolved through the dialectic of Absolute Idea." To this remark Mandel adds "...Marx, in the *Manuscripts,* in no way relies upon any 'idea' but upon the concrete analysis of *social* contradictions; communism, from that moment on, is no longer the result of 'the idea of the development of contradictions' [as Bottigelli had expressed it] but of the *practical struggle of the proletariat.*" [7] The identification of "idea" with "praxis," as we shall see, engenders the necessity of speaking in *concrete* terms when referring to economic or social evolution, and politics. Hence any Marxist "theory" will be programmatic; and the fixity of, for example, Soviet or Chinese political "theory" can only take the form of the ideologies evoked in the opening paragraphs of this chapter. Thus the real cleavage between Marxist Thought and Communist Propaganda is easily exposed. Consequently, we shall be more concerned with the writings of the early Marx, whom everyone seems to recognize as a philosopher, especially since current philosophical interest centers there and is, for the moment, the sole but vital contact between Marxist and Christian thinkers.

Marx was one of the "Young Hegelians," entering the University of Berlin in 1836. Hegel had died there in 1831, leaving his followers to carry on—and carry on they did, soon dividing into a "left" and a "right." Gans, who succeeded Hegel in the Chair of Philosophy in Berlin, was of the "left," liberal Hegelian group, impressed by the democratic progress evidenced by the French Revolution and the socialism of Saint-Simon. This fit well with the interests of Marx, who at an early age was involved with the liberal Kantians and the liberal politicians frequented by his father, all passionately in favor of the principles of the French Revolution. Thus the liberal bourgeois movement in Germany (only partly liberated from its traditional feudalism) rejoined the social-democratic movement attached to Saint-Simon's idea of the necessity of liberating human beings totally through a more just distribution of wealth and a more rational organization of production. Nevertheless, these groups were led to reflect more deeply upon the middle-class takeover of the French Revolution and the increasing antagonism in Germany between the working class and the bourgeoisie. At the university, then, remarks Garaudy, "the young Marx suffered though still confusedly the

[7] Mandel, 170.

ruptures of a world in the process of dying and the contradictions of a world in the process of being born." [8] Goethe was the model for the young Marx of the expression of the contradictions in middle-class humanism: there had been a hope that the race of men could become one through a harmonious division of labor, achieving an organic totality which could, by its very coherence and unity, extend its powers limitlessly in time and space, to conquer the universe. However, within this promethean task, then on its way to becoming a reality, another movement was at work: the individual appropriation of riches and power thus procured rendered the totality of mankind into dispersed and opposing units. "The same regime which opens limitless horizons before mankind." remarks Garaudy, "condemns the greater part of mankind to a beggarly human existence." [9]

The Hegelian Left never agreed with Hegel that Absolute Spirit had accomplished its task and had come to rest in the established order of nineteenth century Germany; for them, and even more for Marx, Hegel's *Phenomenology of Mind* was not a reality but a program. In addition, Marx's own private refusal to admit a possible reconciliation of the Ideal with the Real led him very early to conceive the programmatic of the Hegelian Left as the achievement of history in *praxis*, action, and struggle, and to the conviction that *thought* about reality could not substitute for reality itself.

The basic intuition and the key to Hegelianism is the divinization of man through Absolute knowledge, the reconciliation of the finite and the infinite through religion, science, morality, art, philosophy, and law —the ways in which the world becomes one with reason, and man becomes one with the totality of all being. The young Hegelians dreamed and thought, as Marx later remarked, that "the philosophy of Hegel was the German theory of the French Revolution," since, in their eyes, the conquest of nature through science and technology which was being effected by the forces liberated in France proved the Hegelian Theses; but at the same time the Hegelian system ended with a final reconciliation with the world *as it is*, and "the 'rational' consecration of the bourgeois order, as formerly the theory of 'divine right' had given its consecration to the feudal order and to absolute monarchies." [10]

For this reason, the Young Hegelians could arrange themselves into positions of "left" and "right," depending upon whether they were

[8] Garaudy, *Karl Marx*, 16.
[9] *Ibid.*, 17.
[10] *Ibid.*, 21.

willing to go along with the constituted system of Hegel or whether they wanted to develop the constituting element of the system, the dialectical method, and push on into the future. The divisive element was David Strauss' book, the *Life of Jesus*, which effectively dissociated philosophy from religion, and logic from history. From this point, the Left became atheist, and turned its attentions away from a dialectical understanding of "what is," and began to extend the dialectical movement, which Hegel had stopped in the present, into the future.

Marx himself, from the time of the publication of his doctoral dissertation on the philosophy of nature in Democritus and Epicurus, insisted, against the ancient determinism of the former, that the Epicurean notion of freedom opposed to it was insufficient as long as it did not deepen the consideration of the dialectical relationship between necessity and liberty—and thus the relationships between man and the world, between man and man.

At almost the same time, Feuerbach's *Essence of Christianity*, following the first steps made by Strauss, appeared. Not only did Feuerbach argue against Hegel the incompatibility of philosophy and religion, and of reason and faith, but against Hegel's attempt at total unification by gathering up into Absolute Spirit the totality of the real. In effect, Feuerbach reversed the entire system of Hegel by "putting the predicate in the place of the subject and the subject in the place of the predicate," that is, in other words, by asserting that the idea is a reflection of the world, the world is not the reflection of the idea. Thus the sure establishment of Idealism was forever inhibited.

There seems little reason here to expatiate upon Feuerbach's theory of alienation which followed upon his demonstration of the identification of absolute idealism with religion. Once the absolute idealism of Hegel had been unmasked as a theological apology for the *status quo,* the master ideas of religion which it covered could be understood as projections of the desire of superiority and perfection which man himself produced into a supposed exterior and superior reality. Thus, the essence of man is projected outside of himself and divinized; man is not the creation of God in his image and likeness, but the gods are the creation of man in his own image and likeness. God is the alienation of man, and not, as Hegel imagined, man the alienation of God. Feuerbach then, after liberating man from religion, hoped for a humanistic communism in which man, divided by religion, could be reunited consciously with other men in a materialistic atheism.

Marxist students of Feuerbach see in this inversion of the system of Hegel by Feuerbach merely another system, however. In the first place, Feuerbach merely replaced a dogmatic materialism for a dogmatic idealism, merely naturalized Hegelianism; Hegelian metaphysics became an anthropology, and the idealist Hegelian system, having become a materialist Feuerbachian system, ended also in a theology, deifying love and the dialogue between I and Thou. Secondly, Feuerbach could not help the Young Hegelian Left find what they were looking for: a way to make philosophy an instrument of political and social action; but his effort opened the way toward a philosophical materialism and a dialectical method which would help Marx on his way to political radicalism and philosophical communism. In the fourth of his theses on Feuerbach, Marx, who had already realized that Feuerbach's system merely replaced ideas of "spirit" with ideas of "matter" without solving any concrete problems, and had had enough experience through his editorial career of the urgency of solving those concrete problems, stated flatly that a temporal basis was lacking in Feuerbach's system, and that, lacking a basis in reality, it neither explained nor offered a corrective for the ruptures and the contradictions in our temporal experience. It was necessary, decided Marx, to begin a study precisely of the "temporal basis" of human existence in order to know how to suppress its contradictions.

It was explicitly Moses Hess, together with his own experience as editor of the *Rhineland Gazette*, which urged Marx along the path toward the suppression of alienation, no longer that engendered by religion, as for Feuerbach, but that engendered by private property. Hess' utopian communism was not founded upon the theses of the class struggle and the revolution of the proletariat—that was to come as the contribution of Marx himself—but was in line with the theory of alienation struck out by Feuerbach. It was in this spirit that Marx undertook his critique of Hegel's theory of the state, in 1843. What was, really, the relationship of the state to "civil society" (understood as the totality of economic and social interests)? Marx denounced the philosophy of Hegel for establishing the Prussian state as a "rational" system, the incarnation of absolute Logic, and for thus justifying monarchy and monarchical institutions. If Hegel was correct, it was the state which created and ordered society; so (as fully "rational") its reactionary political institutions and its property owners found their ultimate and divine justification. In his paragraph-by-paragraph refutation of Hegel's theses, Marx could establish, in opposition to Hegel, that it was rather society which created and ordered the state—with the necessary consequence that the Hegelian-

Prussian state was simply the expression of the class-interests within it, hence the political regime guarding the interests of the private property of the few.

More radically still, Marx upheld against Hegel not merely the inversion of the relationships of civil society and state, but those of subject and object. In his famous commentary on paragraph 262 of Hegel's philosophy of the state, Marx noted that Hegel, rather than founding the state upon the family and its property, reversed the relationships, making the latter the product of the state, and remarked that in paragraph 262 "is found the entire mystery of the *Philosophy of Right* and Hegelian philosophy in general." This remark is interpreted as containing the totality of Marx's attack upon Hegel: not merely the creative role of the state with regard to "civil society," but the creative role of the Idea with regard to sensible and temporal reality in general.

Quite early, then, Marx was armed with a concrete philosophical basis for his work, and a concrete, dialectical method; but what was to make it *move* in the concrete? The Fichtean concept of *praxis* was to be given a concrete historical and social redefinition. It was Fichte whose thought had integrated man most fully with the world, in his insistence that man could have no consciousness of himself except through his own construction of the becoming of the world; once Marx liberated his own thinking from its Kantian and Hegelian abstractions, he could, with the help of Fichte, "search for the idea within reality itself," convinced of the unbreakable *unity* of spirit and world.

Auguste Cornu, in his encyclopedic work on Marx and Engels, listed the great ideas which Marx borrowed from classical German philosophy: the union of spiritual and material reality and their reciprocal relationships, the radical immanence of spirit in history, the notion of becoming, and of contradiction as the motive power of that becoming. Garaudy, in his work on Marx, found the Fichtean source of the great Marxist theories of subjectivity, of liberty, and of *praxis*, and remarks upon the common parentage of Marxism and of Existentialism in the Fichtean idea of man the creator; but the Marxists followed more closely Fichte the rationalist, while the Existentialists followed Kierkegaard the irrationalist. For Fichte and Marx essence and existence are in a dialectic relationship: existence, in man, precedes essence; existence is action, creation, and self-creation; what is created is then submissive to its own contradictions. It is more to Marxism than to Existentialism that we are indebted for the axiom that "man is not a given but a task" since human existence is action and creation. Thus, freedom is opposed neither to reason nor to necessity, the history which it creates has always a meaning; and for Marx, the meaning of history limits the freedom of man by the conditions directly given and inherited from the past. It is always out of the contradictions of the present that we must create the future.

From this concept of freedom is derived that of subjectivity; in *The Holy Family*, Marx is careful to oppose the abstract Ego of Fichte to an egoistic individualism; and in *The German Ideology*, the Ego is found as "project," having no sense except in its future-to-be-created. Here again, Marx's notion of subjectivity as project differs from that of some "existentialist" thinkers, as Garaudy remarks; the "nothingness" proclaimed by the latter as liberty is, for Marx, rather non-being: "Existence... is not... the solitary and desperate tête-à-tête of subjectivity and transcendence, but the creative and free act." [11]

Marx's theory of liberty and subjectivity leads naturally, following Fichte against Kant, to a theory of praxis. Kant's ideal of practical reason shows us the individual in struggle between the duties of a solitary conscience and nature; Fichte's ideal of practical reason includes the creative activity of man, with a consciousness of the unity of theory and practice, together with an idea of freedom as "conscious necessity."

This is a very sketchy and incomplete picture of one of the sources of Marx's thinking, that of the German philosophical tradition. The other two, that of English political economy (Adam Smith, Ricardo, and Mill) and French Socialism (Proudhon, Saint-Simon, and Fourier) can only be mentioned. With these, in his *Manuscripts of 1844,* he attempted to bring some order into his thinking, centering his efforts around the notion of work as alienation, toward which his sources all converged.

Marx's question about alienation is that of the relationship of human activity to the objects and to the institutions which it creates. Going back to Hegel's *Phenomenology of Mind*, Marx recognized in the paragraphs on master and servant and on culture the basic theses on man as his own creator, on the becoming of history, and the dialectic involved in creation and becoming. The whole of history is this continuous creation of man by man in a dialectical development: the negation of negation. Unfortunately, Hegel's theses were only abstract, logical, speculative expressions of historical movement; and Marx could wonder as did Kierkegaard, how Hegel could manage to put movement into logic. Hegel's limitations were that he knew *work* only as the abstract work of the spirit, and not the real work of men working to transform nature,

[11] *Ibid.*, 49.

society and themselves, and that, in his view of work, (understood as the essence of man), there was only positivity. Here Marx found the center of all the contradictions of Hegelian philosophy and of classical economics: all refused to see work as alienation, while they (Hegel, Adam Smith, Ricardo) themselves, as thinkers, were abstract forms of alienated man, moving within a system of alienation. They thought and acted and theorized from the bourgeois, necessarily alienated, point of view. Marx's thought was to be simultaneously a critique of classical economics and bourgeois capitalist society, of classical idealism and of positivism; the effort was to be not to change our idea of the world, leaving the world as it is, but to change the world itself in order to change our ideas of it.

This is why Marx chose the point of view of the proletariat; it was necessary to change the point of view from that of the dominating class to that of the dominated. Marxism, remarks Garaudy on this point "is the philosophy of work first of all because it is the philosophy of workers, of those for whom nature is not a creation or an alienation of spirit, but the material of work." He continues his reflection: "From the moment one has understood that philosophical illusions are engendered by the existing world, it is evident that these illusions cannot be dissipated by philosophical critique alone, but by the practical reversal of the social state which has given birth to them." [12] It is easy to see that once "alienation" is considered no longer an *idea* but a *reality*, the reality of alienation must be found in the reality of the worker. Marx found the essential characteristics of this kind of alienation in the alienation of the product of work, which Garaudy names "dispossession;" in the alienation of the act of working ("depersonalization"); in the alienation of generic life ("dehumanization").

The alienation of the products of work takes place through the natural division of labor through merchandising. When put on sale, the product is separated from its producer (as Hegel had remarked already in the *Phenomenology of Mind* regarding the production of ideas), becomes merchandise and escapes from the laws proper to its creation into the impersonal domain of the market-place. By the same token, the sale of the human force which is work becomes merchandise and is entered into the account books of an enterprise on the same footing as the price of raw materials.

This must be inevitable in any society in which there are producers and consumers; but when the institution of "private property" includes

[12] *Ibid.*, 77-78.

the "means of production," then the natural creator, the human worker (slave, serf, or proletarian) is separated from a natural relationship with his creation. The conscious ends of the worker, which he himself can give to his creation, are no longer his own; his own work is no longer the attainment of his own goals and projects, and he attains the ends of another. In his work, then, man ceases to be man to the degree that he no longer assigns his own ends to his work. He himself becomes a means to an end, an intermediary in the objective processes of production, and is dispossessed of the fruits of his labor.

The act of working itself is also alienated, and not only the product of labor, since the means and methods of working are also imposed upon him. His gestures and rythms of working are commanded according to the place he has in the process of production. All his working-activities are planned in advance, and without his being consulted; he becomes "a flesh and blood appendix in a steel machine." The worker is thus depersonalized.

Finally, the alienation of the worker from the product of work and from the act of working through the establishment of the system of private property leads inevitably to the alienation of the generic life of man, his dehumanization. Once the totality of the means of production, of scientific research, of technical means of achieving the ends of a given culture (and the power over society which they represent) are taken over by the few, the very fruit of work and thought which is the patrimony of that culture and society, the creative heritage of the past, passes under the control of the few. The many are thus dispossessed of their history and their humanity.

In the societies motivated by the spirit of private property, according to Marx, the true being of man, his creative activity, stagnates into the morality of "having" rather than "being." For the worker, work and creativity become objectified, become superior to himself and dominate him; for the owner, the worker's work-become-merchandise is a further means of production, thus dead and inhuman, a capital to be possessed. For the Marx of the *Capital,* the more *having* grows in the hand of the possessor, the more the *being* of the creator, the worker, is impoverished. Human relationships are transformed into the relationships between objects, on all levels—economic, political, and spiritual—once man, whose essence is to pursue consciously his own ends and realize them in his work, is alienated through the inhuman laws of having.

It is evident that Marx saw something in the capitalist system which was already leading toward crisis; could he have foreseen, though, the crisis of May 1968, in Paris? Contemporary observers are almost unanimous in calling the event a "cultural revolution" (crushed, to be sure, but revolution nevertheless). Would Marx have understood that something ended there, that history, in a sense, actually came to an end? The Latin Quarter of Paris, which had witnessed eight centuries of student unrest, watched again the young people of Paris stand and fight technocracy and the society it represents—a society of consumers. It was a violent and spontaneous revolt against the positivist world of "facts," against a reality which, for all its measurability, technicity, and objectivity, was, nevertheless, totally absurd. Those who witnessed the event in its depth could only conclude that the young logically, though desperately, drew the conclusions imposed by the demystified and desacralized reorganization of the world by their elders. André Piėttre understood that incoherence and illogic was the fault rather of the elders, who dared not see what is implied in a completely "positive" world of *things*. "Society is only a fact, a technique, a system of production and consumption; so love is only a fact, a technique . . ." After the age of heaven, he concludes, and after the age of man, we have been left with the age of objects; and it was in the name of man, if not of heaven, that the young people of the Latin Quarter built their fragile barricades against the "forces of law and order."

And, of course, the university system, as enmeshed in and at the service of, the society of producers and consumers, had to suffer the same revolution; and so did books themselves (which only teach methods, analyze techniques to study how books are made, as things to be produced and sold, and not what they have to say). The "human sciences" were on trial too, all too clearly dehumanized, and at their summits the prophets who announce the death of man. The event was long in coming, to be sure; the Renaissance had installed a class-humanism without true humanity, the Enlightenment had introduced the religion of man under the aegis of the deification of progress, our own age merely stole the crown and enthroned Science and technolatry. The revolution happened then, as it happened at Berkeley and at Columbia for the same reasons.

Paradoxically, however, it was *not* the proletariat, the working class, which struck out for the "rights of man." The workers at the Renault automobile factory in Paris, who had gone on strike in sympathy with the students, returned to work with the promise of a ten percent increase in salary. The proletariat of Cicero, Illinois, were (as Marx would say) the instrument of oppression of capitalist society, the society of having, when they stoned the black people who arrived in their neighborhood to prospect for a bit of soil and equality. And so on. One can make a long list of the incoherencies of the contestable infrastructure of capitalist society without, for all that, being able to draw Marx's conclusion about the inevitable wasting away of capitalism to the benefit of socialist society and economy.

Space does not permit an investigation in depth of this fascinating phenomenon. A few reflections, however, may not be amiss.

First, is it perhaps possible that the students (in Paris, New York, or Berkeley) were conscious of the alienation present in contemporary society, suffered it most, and yet were economically free enough to exteriorize their revolt against it? Paul Ricoeur has meditated upon the paradox of societies wealthy enough to support their own revolutionaries, with the hope that such revolution, though materially a failure, might yet have a prophetic role in society.

Secondly, is it perhaps possible that the proletariat, alienated though it might be, unconscious though it might be of its alienation, has solved its problem in a way unforeseen by classical Marxist thinking? Instead of overturning the capitalist regime as it was supposed to do, has it turned the system to its own profit? The trade union movements in capitalist countries have achieved the goal of freeing the proletariat from slavery to the machine and, in large measure, to the "Bosses." The working classes, as in the United States, can even participate in the capitalist enterprise by saving enough money to buy stocks in the private enterprise sector. Much work will have to be undertaken in order to decide whether the working classes in capitalist countries have been seduced by the capitalist ideology, or have effected an unconscious revolution of their own; this cannot be done here, but it ought to be remarked that Marxism has never been attempted in any of the countries which Marx predicted it might be. Some have noted that a Marxist experiment might be, perhaps, more meaningful for contemporary industrially developed countries than it was for Russia in 1917.

Third, following from this observation, there is a more basic question involved. It seems that, basically, both capitalist and socialist/communist societies are undergoing the same kind of dynamic. It is very difficult to untangle the elements of this dynamic, but a few leads are available. The basic misunderstanding of Marx, for which capitalist ideologies as well as communist ideologies are responsible, is to imagine that his project was the transform by revolution the power of *having* within the same dynamic of alienation. The strongest impression given by the

Marx of the *Manuscripts of 1844,* however, is the contrary. It is the hope of the realization of the *being of all men* (of the worker alienated from his work as well as the exploiter alienated from his possessions) through the abolition of the alienating structures of a society of social classes. Marxist thinkers and other philosophers who are taking a "grass roots" look at Marx's program accept as axiomatic that the first principle of Marxist morality is participation in the liberation of the total man, as described by Lefebvre in the beginning of this chapter.

Where is the real problem, then? Leaving aside the question of whether hierarchisation is inevitable in *any* society, some reflection upon the actual occurrence of it in given societies seems imperative. As early as 1918 Max Weber predicted the failure of socialism through the *inevitable formation of bureaucracy.* Weber's analysis and critique of bureaucracy was very general and very detailed; it has found echos in later imaginative literature (*The Man in the Grey Flannel Suit* kind of book) and in the sociological theses of Mill's *Power Elite,* following Weber. For Mills, the power elite ensconced in the bureaucratic hierarchy becomes more and more free of control, more and more irresponsible (with a "superior immorality"). Parsons, in opposition to Mills, saw within bureaucracy a dilution of power, and the possibility for the individual to be freer and freer of it, even though it continued to lurk in the background. It is this "background," however, which ought to be brought into foreground. It is all very well for Harvey Cox, the evangelist (inspired also by Parsonism?) to preach the increasing possibilities of individual freedom within the bureaucratic system; that system is still there, still operating, and, as we remarked earlier, seems more and more to run itself. The *true* centers of social and economic decision-making are less and less easily identifiable; each member points to his superior as the locus of responsibility, and when one reaches the top one finds nobody home. This is true in civil society as it is in ecclesiastical society and the "military-industrial complex;" "the system" is as it is, and we really cannot do anything about it. Personal freedom, as Cox and others express it, *may* be increasingly possible; the "freedom" of the whole system, with full and personal participation, is increasingly impossible, and the system becomes increasingly repressive. (To cite only one example, we attempt to "cool off" or "speed up" our economy without really being sure that our efforts will be crowned with success—or whether the resulting inflation, on the one hand, or increased unemployment, on the other hand, can even *be* controlled.)

Much ink has been spilt in capitalist societies worrying this problem;

a review of the plight of Western man and his possible avenues of escape has been undertaken by writers in many contexts: technological (Ferkiss' *Technological Man: the Myth and the Reality),* religious (Hart's *Unfinished Man and the Imagination),* and so on. Kafka's bureaucratic experience has given us a desolate picture of the human condition in *The Trial* and *The Castle,* and these should not be overlooked.

It has not been forgotten that Weber's analysis of bureaucracy was an indictment, in principle, of the very possibility of communist society; nor is it by accident that we have drawn conclusions from it applicable to our capitalist society, against which the contemporary cultural revolution struggles. There is no intention of masking the justice of Weber's prophecy regarding the inevitable bureaucracy in Marxist society.

The most rigorous Marxist writers (we take Mandel as an example) are convinced that "for Marx the phenomenon of alienation is anterior to capitalism. It is linked with the insufficient development of productive forces, with a market economy, with a monetary economy, and the social division of work. As long as these phenomena survive, the survival of a certain form of human alienation is inevitable."[13] The program, then, is the revolutionary or gradual abolition of these phenomena. Given the fact that such phenomena do not seem to be on the way out, these writers must speak always of "*progressive disalienation,*" and contrast it with the theses of the "inevitable alienation within any industrial society" as advanced by non-Marxists. In doing so, they must take into account the *de facto* bureaucratic atrophy in socialist countries, together with the absence of true social democracy in the political sphere, of true worker representation in economic planning, of the absence of creativity in the cultural sphere. To keep eschatological hopes up, then, they must consider the *status quo,* inevitably, as a "period of transition" in the classic stages of progressive disalienation from primitive communalism through slave societies, feudalism, capitalism, to final socialism. They must further interpret contemporary unpleasantnesses (including the Soviet invasion of Czechoslovakia in the Spring of 1968) as regrettable concomitants to what is only a "first phase" in what *must* be the true process of man's history toward final reconciliation.

Marx himself, however, even in the *Capital,* could consider the "Asiatic mode of production" as taking a different course from what is now considered the necessary sequence of events in the passage from a classless society to a class society and back again. Many con-

[13] Mandel, 184-185.

temporary Marxists, seduced or coerced by Stalin's "philosophy of history," attempt to make the "facts" conform the "idea," contrary to the very basis of Marxism, and forge an ideology to support the use of coercion in the application of the idea. In other words, Marx's effort at understanding achieved certain results, and these results have been turned into an ideology and a dogmatism which do not seem to be furthering the evolution of Marx's new and total man. A most fruitful dialogue is now taking place between Marxists and Christians, in an attempt to come to terms with this situation. For the Christian, the question is whether or not the Marxist dream of the total man can be fulfilled in the Christian context, since the Christian dream up to that point is the same; for the Marxist, the question is whether the Christian can shed some light upon the burning question of personal destiny, since Marxist ideology looks forward only to the fulfillment of history and it is not certain that the individual is willing to sacrifice himself for that end.[14]

What now can be said of the oppositions and contradictions within as well as between the great twentieth century ideologies? What can be said has been best expressed, I think, in the passionate statement of Edgar Morin: "The crisis of classical politics brings it close, up to a certain point, to radical politics. The crisis of radical politics brings it close, up to a certain point, to traditional politics. There is still the native opposition between the two politics, one born of a transforming aspiration, the other of a regulating aspiration; but perhaps exactly there begins a possible convergence."[15] Morin, in his genial way, characterizes the "generic man" of Marx as a "promethean, bastard of the Spirit of the Hegelian world and the property playboy of the world of bourgeois humanism," and finds that, in his very essence Marx's generic man is "simple by insufficiency." The idea of man as essentially worker leaves out of consideration (as derived or alienated satellites) imaginary man, playful man, mythological man. In short, Marx's man has no soul, no anguish, no real will to power, no poetry, no folly, no mystery. Morin grants Marx the genius of understanding man as concrete and real, but argues that he omitted a critique of the real which includes imagination. Secondly, Morin wonders why all the Marxist investigations of the

[14] Algis Mikunas, in his paper, "Marxist-Christian Dialogue: Openness and Invitation," soon to be published, analyses in greater detail the motives forcing both groups to dialogue—chiefly the multifaceted challenge of existentialism. He promises a study of the impossibility of a basic opposition between Marxism and Christianity because of their roots in a common metaphysic.

[15] Morin, 15.

reasons for the exploitation of man by man take for granted that, in fact, one group of men *naturally* seeks to exploit another, and that the phenomena which provoke exploitation (scarcity of goods, a market economy, etc.) do not, rather, provoke other forms of social organization which are more cooperative, libertarian, egalitarian, fraternal? Was Marx's hope for progress through class struggle simply founded on a certain historical situation? Does the proletariat really have the weight to swing the balance of history in the right direction? Is Marx's urban proletariat different from any other present or past oppressed group? "Did Marx," he asks, "charge the proletariat with a practical hope or a messianic dream?" Without messianic hope, he continues, "Marxism would be, like Freudianism, a passionate diagnostic, a tragic optimism, a pessimism only compensated, though profoundly compensated, by the idea of progress."[16]

We have noted that history has not properly followed the script written by Marxist ideology; and it is evident that nowhere on earth has the proletariat fulfilled its historic mission. Marxist ideologues are still awaiting the dialectico-historical parousia. Are we to wait with them for the inexorable laws of political economy to fulfill the dream of the total man? And can we cast easily aside, as they have done, the thesis of the "inevitable alienation within any industrial society," which they take to be only special pleading on the part of apologists for the capitalist *status quo?*

Philosophically speaking, Morin goes deeper in his "Foundations for a Politics of Man." Remarking, as many have done, that Freud lacks *homo faber,* and that Marx lacks a *psyche,* he thinks that these two views of man must still be reunited if the necessary modern revolution, the radical change in human relationships and the realization of the total man, is to succeed. Some efforts have already been made to reconcile or reunite these two views. The French journal, *Man and Society* (*L'homme et la société*), subtitled "an international review for sociological research and synthesis," devoted its January, February, March, 1969, issue entirely to the problem of "Freudo-Marxism and the sociology of alienation." The question was raised by the now-historic controversy between Marcuse and Fromm. Boris Fraenkel, the editor, saw a convergence of opinion among all the contributors in their radical critique of alienated society, and some complementarity in the basic concepts of dialectical structure and historicity. The same re-emergence

[16] *Ibid.,* 22.

of the Freud-Marx question is symptomized by the almost weekly publication of spates of new books on one or the other, or both, authors. The reader can profitably follow through the dialectic of these two basic views (let us say the archaeology of Freud and the eschatology of Marx) and attempt to forecast their eventual reconciliation. It seems to the author, however, that Marxist thinkers will have to give up, in the long run, their contradictory stance which insists that society and history will advance only through conflict, but achieve a final reconciliation, once they admit the unresolvable conflict within the interior dialectic of man himself, the unresolveable dialectical confict of individual human existence.

Morin's own effort is to go beyond even these basic conflicts to a more generalized statement of what he calls "anthropolitics" and the possible conjugation of the themes already reviewed in these pages:

> The revolution is in crisis in the world. But at the same moment, it is the world which is in revolutionary crisis. At the very moment we toll the bell for the death of the revolution, we must toll it as well for conservatism and evolutionism. For if we are not living the revolution announced by the revolutionaries, we are living the most fantastic revolution in the history of mankind. *We are not living the revolution of civilization; we are living the savage revolution* provoked, conducted, and accelerated by science. Savage because lacking guiding ideas and regulations, escaping all control—beginning with men of science who control as little the cause and the movement of their discoveries as they do their exploitation....
>
> The scientific revolution, which has its resonance in the technological revolution, and then more and more disturbs, modifies, and transforms the entire social body, is an unchained revolution.
>
> This revolution, first oriented toward the natural milieu, has approached the human being in concentric circles, enveloping him, squeezing him, penetrating brusquely into its soul by the flux of mass-media, preparing to penetrate the genetic arcanum, the holy of holies where precisely sits the power of human biological transformation.[17]

Recent technological and scientific advances make it clear that we can do, materially speaking, anything we can imagine, whether we imagine a utopian end to history or the biological transformation of individual human beings (for the former, we need only consult *Walden Two* or the newspapers which analyzed the dynamics of the Russian invasion of Czechoslovakia; for the latter, *today's* newspaper is our source). The time lapse between imagination and realization becomes shorter and shorter. The only question not yet solved is, again, whether we *ought* to carry out whatever our scientific and technological imagination can conceive—and here "scientific and technological" has its fullest extension of meaning, including the total social and economic and biological

[17] *Ibid.*, 29.

organization of humankind. Are we to build an "ought" upon the "may be," though we have refused to build it upon the "is?"[18]

If we define "the revolution" as the radical change in human relationships, as an ambition which goes beyond the abolition of exploitation, we can conclude with Morin, that "The revolution is liberating, but its end is not liberty. Absolute liberty comes to mean the disintegration of all social bonds.... It is the absolute negativity which would have nothing further to negate; so it would deny itself insofar as it is negativity."[19] *Absolute* freedom, then, would be a slavery to the unconscious, and the illusion which Freud named it. We are left, then, with the relative autonomy which Freud and Ricoeur have so well analyzed. But the very relativity of our freedom is rooted in the historical situations of the two main opposing political-social-economic ideologies in today's world—both bankrupt, both sterile and sterilizing—and more deeply still in our own inner being.

Morin's book begins with a realization that politics is in pieces; but more and more thinkers are coming to the conceptualization of a world in pieces, a conceptualization of our fragmented lived experience. The general feeling is that we have now arrived at the Iron Age of the planetary era, and out of the present chaos we must attempt to build not the twentieth century but the twenty-first, if only to save the twentieth. The general conviction is that, in order to do so, we shall have to learn an ethics of imagination—but that is another story, and another book.

[18] In the dominant empiricist philosophical tradition, no ethically imperative statement can be deduced from statements of fact.

[19] Morin, 33.

SELECTED ADDITIONAL READINGS

FOR CHAPTER V

Aron, Raymond, *Marxism and the Existentialists,* New York, 1969.

Bauman, Zygmunt, "Modern Times, Modern Marxism," *Social Research,* 1967 (34), 399-415.

Bigo, Pierre, "The Enigma of Marxism," *International Philosophical Quarterly,* 1965 (5), 637-646.

Birnbaum, Norman, "The Crisis in Marxist Sociology," *Social Research,* 1968 (35), 348-380.

Blanchette, Oliva, "The International Symposium on Marx and the Western World," *International Philosophical Quarterly,* 1967 (7), 129-137.

Borchert, Donald Marvin, "Marx, Social Change, and Humanization," *Zygon: Journal of Science and Religion,* 1969 (4), 169-187.

Easton, Lloyd D., and Kurt H. Guddat, eds., *Writings of the Young Marx on Philosophy and Society,* New York, 1967.

Feinberg, Gerald, *The Promotheus Project: Mankind's Search for Long-Range Goals,* New York, 1968.

Ferkiss, Victor C., *Technological Man: The Myth and The Reality,* New York, 1969.

Fromm, Erich, *Marx's Concept of Man,* New York, 1967.

Garaudy, Roger, and Quentin Lauer, *A Christian-Communist Dialogue,* New York, 1968.

Gregor, A. James, "Marxism and Ethics: A Methodological Inquiry," *Philosophy and Phenomenological Research,* 1968 (28), 368-384.

Kolakowski, Leszek, *Toward a Marxist Humanism: Essays on the Left Today,* New York, 1968.

Lessing, Arthur, "Marxist Existentialism," *The Review of Metaphysics,* 1967 (20), 461-481.

McHale, John, *The Future of The Future,* New York, 1969.

McLellan, David, *Marx Before Marxism,* New York, 1970.

Petrovič, Gajo, "Marx's Theory of Alienation," *Philosophy and Phenomenological Research,* 1963 (23), 419-426.

Ricoeur, Paul, *Fallible Man,* trsl. Charles Kelbley, Chicago, 1965.

—, *History and Truth,* trsl. Charles Kelbley, Evanston, Ill., 1965.

Schaff, Adam, *A Philosophy of Man,* New York, 1968.

Sommerville, John, "Marxist Ethics, Determinism, and Freedom," *Philosophy and Phenomenological Research,* 1967 (28) 17-23.

CHAPTER VI

THE STRUCTURALIST MODEL

MAN THE SOURCE OR MAN THE PRODUCT?

As our investigation proceeded, the models under review became less and less "neat;" yet, in the midst of the diversity of applications of theories, a *fil conducteur* can stil be found which unites them and which makes, at the same time, the opposition of world-views more and more apparent. The most striking model is the "structuralist" one. On the basis of the same data concerning the nature and processes of man, culture, and society, structuralism (which is proposed as a method, yet used as a philosophy or even ideology) has invaded most of the human sciences.

Structuralism is, then, primordially a method of analysis—the analysis of structures. Basically, a "structure" is an autonomous entity characterized by internal dependencies. This all-too-gnomic a statement has been lucidly explicated by Piaget as a system of transformations which, *as system,* implies a lawfulness of organization independent of the elements which compose it. A system is characterized, first, by its being a *totality,* that is, whatever the "composing elements" in the system, they are subordinated to the laws which define the system as system. Second, a system is characterized by multiple *transformations,* interdependent with each other and with the totality, that is, they are dependent upon the structure itself. This points to a third characteristic of structure, that it is *self-regulating,* tends toward the conservation and enhancement of the system itself, on the one hand, and closedness toward all other systems on the other.

The method of analysis which is structuralism, as it has heretofore been employed in the human sciences, has as its aim precisely the definition of structures (defined by inner relatedness) and the laws of their transformation. The structuralist philosophy-ideology which has grown out of this methodology has asked the question, "Is man, his culture, his society, and all his works, basically fixed and static (since structurally identical throughout history), or is there at least a hint of process, not purely epiphenomenal, which is due to the creativity of individuals?" Put more simply and declaratively, the structuralist philosophy-ideology maintains a contrary position to that of the phenomenological one which we have been considering, in which human consciousness is considered the origin of structure and meaning, and maintains instead that "human being" is no more than a phase in a larger and all-encompassing structure (a determined self-regulating totality governed solely by internal transformational rules). The phenomenologist would contend that human being in the world is the converging, or radiating, center of meaningfulness, while the structuralist would reply that this "center" is elsewhere or nowhere, and that the creative role of human being within the system is illusory.

It is impossible to say, at present, whether structuralism is primarily an anthropological, linguistic, psychoanalytic, sociological or philosophical phenomenon. Its adherents in all these fields insist, like phenomenologists, that it is primarily a *method*; but as methodological, it encompasses a number of philosophical implications which, in their explicitation through methodology (as in the case of phenomenology) show a coherent though contestable image of man. This image is expressed in the terms cited above; and the purpose of this chapter is to follow the grand lines of the formation of this image. The beginnings lie in the alliance of sociology with linguistics. Its endings are not yet in sight; there is at the present time already a "second wave" of structuralist studies which may put some order into the disorderly house created by the battles of the "humanists" with the "structuralists."

It seems incontestable that Emile Durkheim is at the source of contemporary structuralism; for it was he who first conceived the idea of sociology as a science, in his *Rules of the Sociological Method,* a science in its own right, distinct from anthropology and psychology.[1] (We shall see that the latter two gobbled up the enterprise.)

The establishment of the social fact by Durkheim, with its recognition that language could be brought under that heading, was the cause of the linguistic conversion of Ferdinand de Saussure, who until his reading of Durkheim had been noted for his work in the earlier kind of linguistic study, historical linguistics. Thus de Saussure bridged the gap between what are today called "synchronic" studies and those which are called "dia-

[1] See Chapter IV.

chronic" the latter studying the changes (phonetic, grammatical, etc.) which take place in a language over time, the former concerned with actual structures at a given time. Synchronic studies are, of course, essential for good diachronic studies; the scholar must be able to compare at least two "horizontal cuts" through a changing stream of history in order to show what changes have occurred between the two. However, contemporary proponents of synchronic studies seem to find *no structural change* at all between the two moments of research, and conclude that "history" in any sense is impossible. We shall have to return to these considerations; for the moment, we are concerned with de Saussure as the "founder" of structural linguistics as we know them. His *Course in General Linguistics* (published posthumously by his students at Geneva, in striking parallel with the similar treatment of the publication of Mead's *Mind, Self and Society*) shows first his objections against the restriction of the science of language to its historical aspects by his predecessors, and his own program for a "real" scientific approach. Inspired by Durkheim, he hoped to make the synchronic study of language scientific by first identifying linguistic facts on the model of social facts.

The first thing to do, evidently, was to define language in such a way that it could be studied scientifically as an object, without reference to its historical development, and thus independent of individual manifestations, as Durkheim had prescribed.

It is well known that the important aspects of language which de Saussure selected for study were, first, *la parole* ("speaking"), which is evidently an individual manifestation of the common language, and in large measure a phenomenon of free choice:

> it is the sum of what people say, including individual constructions that are the consequence of the speaker's choice, required to produce these constructions.[2]

Second, while speaking is not a social fact, since it is neither general in the community nor does it exercise constraint upon the community, *le langage* (the sum of the acts of speaking within the language community together with the realization of all the rules of the language) enjoys at least a certain generality, being the total sum of linguistic acts, and a certain aura of constraint, being formed under the dominance of the sum of grammatical rules through which communication is possible; but it is still not the social fact required for linguistic study. The trouble

[2] Ferdinand de Saussure, *Cours de linguistique générale*, Paris, 1915. (Translated in Francis P. Dinneen, *An Introduction to General Linguistics*, New York, 1967, 196.)

with *le langage* is that it does not enjoy a total scientific unity, precisely because it is an actual *sum,* a manifold full of the unpredictableness of individual speakers. Third, the true linguistic-social fact, then, is *la langue* itself ("*le langage* minus *la parole*"), *le langage* minus all the individual differences which constitute it as a sum. In short, it is the language as spoken in a community "independent of its individual manifestations." *La langue* for a given language community is the set of sounds taught and learned generally in that community, in accordance with the set of grammatical rules which the community likewise teaches and learns. Thus, *la langue* is the true social-linguistic fact; it is general throughout the community, and exercises constraint over the members of that community. Dinneen explains the differences very nicely:

> Because it is individual, active, and voluntary, *la parole* is not a social fact; *le langage* includes both social and individual aspects and is not, therefore, a pure social fact; *la langue* is the social fact, being general throughout the community and exercising restraint over the individual speakers. Like Durkheim's collective consciousness, *la langue* is not found complete and perfect in any individual. *La parole* includes anything a speaker *might* say as well as the constraints that prevent him from saying anything ungrammatical; *la langue* contains the negative limits on what a speaker *must* say if he is to speak a particular language grammatically.[3]

It is taken for granted that, in order to be scientific, linguistics must be preoccupied with the set of linguistic signs and their organization, not with speech. This does not mean, however, that language is a *physical* fact (quite the contrary; it is speech which is physical); it is a *social* fact—on the model of the fixity of physical facts, to be sure. Dinneen sums up, again very succinctly, the characteristics of this kind of fact:

> It (*la langue*) is a social fact, which can be successfully studied when we consider the pattern behind individual utterances. It is the pattern that is stable, both through time and in the consciousness of the speakers. The pattern changes so slowly that we can even abstract from the reality of the change. Successful assimilation of dead languages should be proof enough that speech is not required of us to master *la langue*.[4]

Let us follow, for the moment, the accumulation of ideas in structural linguistics. It will permit us to rejoin some of the main currents of thought already presented in this book as the major "confrontations" on the contemporary scene. For example, one of the great names in the history of structural linguistics is that of Franz Boas, who was born

[3] Dinneen, 197-198.
[4] *Ibid.*, 199.

in 1858, the same year as Durkheim. Instead of being a sociologist, as Durkheim was, Boas was primarily an anthropologist, and a linguist by avocation. His study of American Indians produced the conviction that it was not the environmental situation of a people which influences the formation of society, but rather its cultural traditions, its language and literature. Boas' work as an anthropologist and linguist is not of primary importance here; it is his influence upon Edward Sapir, who was only twenty years of age when he first met Boas. Sapir gave up his pursuit of classical philology and began, in the footsteps of Boas, to study especially American Indian languages. He soon became aware of the samenesses and differences in languages and dialects. Sapir was the student who first tackled the problem which was left to linguistics by de Saussure: that although *la langue* cannot be changed by *la parole,* the idiosyncracies of individual speakers, it obviously must be so changed. Sapir's response was that change in language results from the differences in what speakers say, since language really exists only in use, although theoretically it is a fixed system; and since language cannot, as de Saussure pointed out, be the *sum* of what people say, the change of language cannot be credited to them personally either, although it might be credited to a general "drift" in the given language as a result of the random differences which contribute to that drift (other idiosyncracies being discounted as non-contributory).

It is commonplace, after de Saussure, to assert that no individual speaker is aware of the change taking place in his language; Durkheim's thesis of the constraint of language upon speakers (even when they feel no constraint) had been upheld. Sapir pointed out that the direction of change in any language, which he called its "slope," could be detected by an examination of the *hesitations* which native speakers make in choosing proper forms of speech. "This drift," says Dinneen, "is not haphazard, nor the consequence of lack of system, but is, rather, the stable dynamism of a developing, living use of language . . ."[5]

We have in hand, then, two large questions to worry: first, there is some connection between language and culture; second, that language in a culture changes. As to the first question, we are indebted to Sapir, especially, for the solid conviction that there are striking parallels between language and culture; in fact, he pointed out, language never exists apart from culture ("the socially inherited assemblage of practices and beliefs that determine the texture of our lives"[6]). On the other

[5] *Ibid.,* 233.
[6] Edward Sapir, *Language: An Introduction to The Study of Speech,* New York, 1921, 207.

hand, there is little, if any, connection between language and environment, except for the obvious enrichment or impoverishment of vocabulary. Another important contribution to the question of the connection between language and culture (defined even more simply as "what a society does and thinks" [7]) is found in the Sapir-Whorf hypothesis, a contribution of Sapir and his brilliant pupil, Benjamin Lee Whorf: that language determines culture through the contents of the concepts which make up the "perceptual world" of the people who partake of that culture. That is to say that, basically, the things in which the people are interested constitute a world interpretable by a language which constitutes a culture. It is this interest which accounts for the kind of "drift" which the changing language will take. Parallel to this is the conviction that experiencing cannot occur except in the context of the formative guidance of linguistic habits in a community of speakers; the world we live in is "to a large extent unconsciously built up on the language habits of the group."[8] According to the Sapir-Whorf hypothesis, then, "men are not led by their experience to the same picture of the universe unless their language backgrounds are similar." [9] Sapir was also convinced, however, that "the latent content of all languages is the same—the intuitive *science* of experience."[10] These two convictions seem incompatible, but in order to understand the direction structuralism actually took in choosing between these seemingly incompatible statements rather than reconciling them, let us draw a few tentative conclusions from the discussion of the first of the worrisome questions mentioned above, namely, the relation between language and culture. Sapir and Whorf corrected the mistake of the traditional grammarian, which was "to deal with all languages in terms of signification, the sole positive fact of language, ignoring the difference in content, which is the result of linguistic value, which in turn is an expression of the structural relations of a term in its system." [11] Dinneen sees that his distinctions between "signification," "content," and "value" help to uncover some of the incoherencies even in the best of structural work: that languages are incomparable (de Saussure), or that languages, being unique structures, "either help or hinder their speakers in making certain observations or in perceiving certain relations," with the result

[7] *Ibid.*, 219.

[8] Edward Sapir, *Selected Writings in Language, Culture and Personality*, Berkeley, Cal., 1949, 162.

[9] Dinneen, 236.

[10] Sapir, *Language*, 218.

[11] Dinneen, 218.

that Whorf and his Hopi Indian friend could not discuss "the same world." [12]

Thus we are left with a certain pluralism; the diversity of languages and the diversity of life-worlds which these languages translate and through which men communicate; but does Dinneen's conjugation of "signification," "content," and "value" serve in a unifying view of the human sciences and the meaning of man? The central difficulty of language, after all, is the problem of meaning and the relationship of linguistic form to it. The philosopher may see in this pluralism simply the atomization of meanings, the unbridgeable gulf between cultural life-worlds, an infinity of perspectives which disparate languages reflect. The problem is enormous; but did the study of language as a science minimize or maximize meaningful communication? Why was the Sapir-Whorf hypothesis concerning the function of content in language not capitalized on? Or, in plainer terms, why did linguistics take the road of *form* rather than that of *meaning*?

One of the answers is that Sapir's elegant structural methodology did a good job in describing linguistic form; "since form alone remains constant in actual utterances, while meaning varies with attention, the object of an autonomous linguistics for Sapir was to be linguistic form." [13] This procedure advanced the synchronic study of language as a science, but showed a certain weakness, later identified by Claude Levi-Strauss (about whom more in following pages) as the purely logical and categorical Kantian legacy left to the structuralist enterprise by Durkheim: In Durkheim's theory, there is a contradiction between the logical Kantianism involved in interpreting the social fact *as fact* and the intervention of a collective symbolism involved in positing a collective consciousness. If the linguistic fact was to be dealt with in terms of form, the unwieldiness of "symbol" had to be eliminated. Lévi-Strauss himself was to insist upon the dialectical character of unconscious psychic structures in their relationship to "objective" systems of ideas. His explanation is a more refined and complicated one, but no less static, as we shall see.

The "weakness" in the structuralist theory begun by Durkheim (i.e. "mentalism") may or may not have been understood in the same way by Lévi-Strauss and by Bloomfield. It is certain, however, that the latter attempted to correct whatever weakness there was in the theory. It is also certain that Bloomfield's intention was to make linguistics both

[12] *Ibid.*
[13] *Ibid.*, 239.

scientific and autonomous, and that he had no intention of letting linguistics be taken over by another science. Perhaps, as Dinneen attempts to show, Bloomfield only wished to demonstrate how meanings could be empirically stated by another discipline than his own. The fact of the matter is that the classical American work, the "bible of linguistics," Bloomfield's *Language*, has all the characteristics of Watsonian behaviorism (i.e. the psychology of Wundt, with "consciousness" purged from it since Watson thought "consciousness" a scientifically indefensible notion substituted for the even less defensible notion of "soul"). Speech, then, and thinking, on the Watsonian model, were considered forms of behavior, understood in terms of the empirically verifiable "response" to a "stimulus." Bloomfield's influence upon linguistic studies, and upon the whole structuralist effort, has been enormous not only in the United States but in Europe, especially among those who, like himself and Watson, are opposed to "mentalism" and in favor of "facts" for purposes of classification and description. Though there have been recent efforts at explanation, language nowadays is thought of almost universally as a code. Even the genius of Sapir, as of de Saussure before him, is supposedly tainted with psychological dualism; only a "*non-mentalist*" (i.e. mechanist or behaviorist) approach to the core-problem of linguistics, meaning, could be acceptable. This was Bloomfield's contribution.

Bloomfield's descriptive linguistics was surely a *tour de force*, and contributed much to his field; and it is certain that the linguist can most profitably do his job if he understands language to be a code, or pattern, or structure, and is not concerned with its psychological or behavioral foundation. Formal studies of language thrive best in this atmosphere, as do purely formal studies of anything. But what of the traditional grammarian's concern for the *signification* of language? In putting the question in this fashion, we are obliquely attacking the second of the worrisome problems mentioned above: the problem of the *de facto* change in language. To elucidate the question it will perhaps be most useful to abandon the pursuit of the history of the purely linguistic aspect of structuralist studies and to consider the wider applications of structuralism by following the passage of structural linguistics back into its origins, and to consider the contemporary structuralist endeavor of Lévi-Strauss.

The project is by no means fanciful; perhaps the most significant thing about the career of Lévi-Strauss is the influence upon it of the work of Roman Jakobson. Jakobson's work was principally a postwar synthesis of the American Bloomfieldian linguistics and that of the "Prague School" which had been out of touch with each other during World War II and had developed along somewhat different lines. In America, Jakobson's synthesis has been incorporated into the new work of Noam Chomsky —the current fad in American linguistics.

Chomsky seems to have become disenchanted with the legacy of Bloomfieldian linguistics and its empiricist foundation, but it is difficult to see how his "transformationalism" finally differs from the "structural linguistics" he had set out to oppose. Chomsky's notion of the *object* of structural linguistics is that it deals only initially with a finite set of sentences actually pronounced by the speaking community (this object strikingly resembles de Saussure's *language*). In opposition to Harris, who maintained that the description of *langage* is the sole objective of structural linguistics, Chomsky maintained that the objective of structuralist description is rather a representation of the *competence of the speakers*. He thus hoped if possible to "break through" the finitude of the manifold of utterances, establishing the possibility that the individual speaker might be capable of uttering an infinite number of sentences; in fact, if the speakers' competence is the objective of structural linguistic study, whether the actual set of utterances is finite or infinite is eventually of no real consequence. Chomsky did admit the possible existence of a finite *corpus*, to be sure; but only as the initial "given" with which the linguist works. If that corpus is adequately defined, the real objective is attained: behind the actual utterances, the power of uttering. This power of uttering, he maintained, is the intuitive knowledge which speakers have of their own language. Here, Chomsky was attempting to undo the negative attitude which American structural linguists, after Bloomfield, had had for "linguistic feeling."

Another aspect of the American structural linguistics of the first half of the century which Chomsky opposed was its pretense of the ultimate utility of classifying the finite manifold of utterances. Since a structure is defined by a closed system of transformations, no information exterior to the system could be used to classify ultimate constituents of discourse. The only solution, then, was to group together elements of discourse which occupy the same relative position in a sentence. A "syntagmatic" classification is finally achieved, in which all linguistic elements are distributed according to a set of possible contexts. A "paradigmatic" redistribution is then undertaken in order, hopefully, to achieve the same kind of hierarchical classification as that of zoological taxonomy. Against this attempt at classification, Chomsky objected that a more ambitious program of explanation ought to be attempted, even if a perfectly organized system of paradigms could be achieved. The explanation ought to bear upon *how* linguistic elements thus classified were combined in the first place.

It was at this point that Chomsky made his most original contribution to structural linguistics: the thesis that classical structural linguistics was dealing only with *superficial* structures, whereas only an analysis of *deep* structures could reveal what was really important, namely, how speakers interpret and judge utterances. Chomsky's effort at showing that and how superficial structures are derived by transformational or generative procedures from the deeper structures of the human speaker led naturally to the hypothesis of "linguistic universals," an innate language-facility which permits speakers to learn and use their mother tongue.

However, more than one scholar has expressed astonishment that Chomsky, in spite of his anti-behaviorist intentions, seems to have taken a path parallel to that of the Bloomfieldian linguists and has eventually caught up with them on their own ground. His "generative" grammer is intentionally opposed to the "distributional" grammer of his adversaries; but it is not at all certain that he has not been seduced by the metaphor of the "generative" process. The final explanation of creativity in language seems to have escaped him—and how could it be otherwise when "structure" is the key term, and when structure must be defined as "system?" [14] It is interesting to note that the most "in" twentieth century linguistic study in America has returned to a kind of theory reminiscent of the classical theories of the Middle Ages, which supposed a standard and universal language (which they presumed to be Hebrew, since that is the language God speaks) from which all actual languages diverge as corruptions due to the unfortunate occurrences at the Tower of Babel. It is assumed, after Chomsky, that unless a child already has a knowledge of language in some sense he could not learn a *particular* language; and the knowledge of "language in some sense" is the unconscous structure of the mind, a universal grammer. Chomsky says explicitly:

> A theory of linguistic structure that aims for explanatory accuracy incorporates an account of linguistic universals, and attributes tacit knowledge of these universals to the child. It proposes, then, that the child approaches the data with the presumption that they are drawn from a language of a certain antecedently well-defined type, his problem being to determine which of the (humanly) possible languages is that of the community in which he is placed. Language learning would be impossible unless this were the case.[15]

[14] Scholarly reflection at this point has centered upon the ambiguities in Chomsky's *Cartesian Linguistics*, New York, 1967.

[15] Noam Chomsky, *Aspects of the Theory of Syntax*, Boston 1965, 27.

Reading the history of linguistics back into anthropology and ethnology, Lévi-Strauss, in France, has translated the accumulated wisdom of structural liguistics back into cultural terms and reinforced the notion of the static character not only of language, but of mind, self, and society as well. The reader is encouraged to explore for himself the way in which he conjugates Freud, Marx, and geology, his "three mistresses" into a universal and coherent theory. A relatively simply example of the model for structural research in ethnology is available in an article by Sanche de Gramont in the *New York Times* for 28 January, 1968. "Imagine," says de Gramont, "someone who has never seen a playing card watching a rubber of bridge. By observing the way the cards are played, he should be able to reconstruct, not only the rules (or structure) of bridge, but the composition (structure) of the deck of cards."

Lévi-Strauss has observed how marriages are arranged in primitive societies, what kinds of myths are told, how gifts are given and received, how cooking is done. The basic patterns of society are reduced to recurring types of operations within the society studied. Thus, says Sanche de Gramont, there are in structuralism two ambitious implications: that the human sciences can attain the rigor and detachment of the natural sciences (as they have attempted to do since the time of Wundt), and that human behavior (now the only proper object of the human sciences) is governed by the limitations of man's cerebral processes. Admirers view Lévi-Strauss' work, continues de Gramont:

> as the final panel of a tryptich entitled 'contemporary Western Thought,' the other two panels being occupied by Marx and Freud.... Lévi-Strauss" contribution to the tryptich is a theory of how the human mind works. Primitive man, in organizing himself into social groups, passes from a natural to a cultural state.... Structuralism postulates that in achieving this passage from nature to culture, man obeys laws he did not invent. These laws are inherent in human nature which is everywhere identical since it is no more than the mechanism of the human brain.... With Lévi-Strauss, the whole humanist tradition goes down the drain. Instead of a free spirit, responsible for his actions, we have a man responding to programmed circuits called structures. The whole body of Western thought, from Plato through Descartes to Sartre also goes into the garbage can.

More important, history goes down the drain too (since it was ruled out in the beginning of structuralism, but now seems to have been impossible in the first place). Finally, says de Gramont, "structuralism is a specific attempt to discredit Jean-Paul Sartre as an outdated thinker and relegate existentialism to the philosophical garbage can."

Structuralism rose out of the amalgam of psychology, anthropology, and sociology, passed through the science of linguistics, and has passed back into psychology, psychoanalysis, anthropology, and sociology, enriched by the incorporation of Watsonian behaviorism—which everyone has been trying (unsuccessfully perhaps because not seriously) to push back into its grave for thirty years. Structuralism has been taken over as the basic explanatory principle of any number of human sciences. A recent volume of the *Yale French Studies* devoted to structuralism contains articles applying structural methodology not only to the study of linguistics and anthropology, but also to art, psychiatry, and literature. The editor comments that "scholars are now trying to lay the basis for a science of signs—semiotics—which would include not only... languages but also any system of signs." In fact, all the elements of human culture can thus be included, since structure, or system, marks the emergence of man from a state of nature to that of culture. Thus, all social and cultural phenomena can be treated as instances of structure, and, it should be noted, of an invariable structure.

A significant recent attempt at structuralist generalization is a book, "Words and Things," (*Les mots et les choses*) by Michel Foucault, which attempts to show that thought originates with words rather than the other way around, and further to show, on this basis, that "man" is simply a product of the language of 17th century humanists, and thus simply a wrinkle on the surface of the natural history of the universe, nothing but a passing fancy of Nature, a particular example of a more universal structure. "Man" will soon, according to Foucault, disappear into the general matrix of the universe which threw "him" up to rot upon the banks of an illusory "cosmos." Another series of efforts by Louis Althusser to apply structuralist method in analyzing man and society from a Marxist point of view met with less acceptance and achieved less success both in the Marxist camp and in that of the phenomenologists. But Althusser's failure achieved the salutary effect of waking up the intellectual world to the fact that structuralist method, after achieving notable success as a *formal* study, had been turned into a philosophical *system* and used as a hermeneutic. This has also become evident in the work of Lacan, as we shall note in passing, even though Lacan's psychoanalytic structuralism is an attempt to make Freud's theories "scientific."

There are, at present two main currents working against the structuralist "takeover." One is an anti-structuralist polemic; the other, more moderate, is sensitive to the positive gains in the structuralist enterprise but simply outraged at its ideological universalization into a total theory about Man. Let us attempt briefly to clarify the point of conflict, and to mention even more briefly the chief protagonists in the debate.

What marks the emergence of *homo sapiens sapiens* upon the planet earth? Thanks to a century of careful work by paleontologists and anthropologists, we know that what is strictly human upon this earth is *system*. Toolmaking, for example, was earlier thought to be the sign of human emergence from the waves of mineral and organic evolution, but now we know that there were pre-human beings on earth who manufactured tools (i.e. *made* implements for specific purposes and did not just use materials lying around them to strike, to lift, and so forth). The difference between them and *homo sapiens sapiens* is that the latter created a *system* of toolmaking. Likewise, in the case of settled, civilized human existence, the passage of man from a state of nature to that of culture is marked by an economic *system*, a system of production and exchange, of cooking, story-telling, and so on. The same can be said for institutions of culture and society, whether the basis of them be religiously structured, as Durkheim insisted, or not. And if we consider the contribution of linguistics to the study of Man, it is likewise clear that a *system of signs* is the hallmark of human speaking. De Saussure long ago pointed out that there is, anatomically speaking, no determining factor which demands that we phonate in order to communicate, any more than there is a determining factor in the "signified" thing which induces phonation, but that a system of signs of some kind is the basis of our possibility of communicating with one another at all.

Now, what ever else system may connotate, it is a system marked by internal interrelationships, and thus structure. But what structure? Lévi-Strauss, together with Jakobson, postulates that it is the determining structure of the human cerebral cortex which not merely limits the possibilities of organization but determines *what* structure will be invariable in our economic, political, institutional, linguistic existence. Differences between "primitive" and "civilized" peolpes is, then, only epiphenomenal, the eternal repetition of constant structures.

Lévi-Strauss has had the greatest influence upon the development of structuralist studies, mainly because of the impressive coherence of his results and the unbending rigor of his methods. His explanation of the emergence of man from the state of nature to that of culture is that this emergence is coincident (in a very complicated way which cannot be detailed here) with the prohibition against incest. This prohibition, however, is not the result of biological observation on the part of emergent man nor the supposed taboo-system imposed by the totemic religion which characterized the most "primitive" peoples, but was rather the institution of a *system of exchange* in which the maternal uncle gives his sister in marriage to another man, in return for partnership in the enterprise of hunting, for example. The basis of all social living thus becomes a *system of reciprocal exchange.* Basically, what is exchanged is the woman, understood functionally in the system as the *sign* of reciprocity. Yvan Simonis, in his careful and faithful explicitation of Lévi-Strauss' structural anthropology, notes the close connection between the latter's ethnological studies and the linguistic theory which underlies them as method, and remarks that the way was open to the master to theorize upon the *reciprocal exchange of values* in communication (given Lévi-Strauss' explicit recognition of the personhood of the woman exchanged). The question was: what kind of values were/are reciprocally exchanged within society? Since it is assumed that the woman is the sign of this reciprocal exchange, it was possible to answer, on the basis of the human value of the woman herself, that what is reciprocally exchanged is of really human value—in commercial exchange, in marriage and family relationships, in discourse and poetry. But if what is actually exchanged is only a "sign," then there is, in structural terms, no reason to attribute any particular "value" to what is exchanged at all—except the purely formal value of the sign as sign. Lévi-Strauss' option to study communication by the "sign" of the woman closed the gate to a possible interpretation of human social life as an exchange of *human values* and imposed the interpretation of social living as a pure *function*, based upon an exchange of *objects.*

A similar use of the results of purely formal studies as explanatory theories not only in anthropology and linguistics but also in biology, paleontology, and so forth, leads to the conclusion that the rooting of structured "human" existence in the forward sweep of the evolutionary process indicates a kind of continuity between "man" and "nature" which can make a book like Desmond Morris' *The Naked Ape* seem quite plausible.

This, then, is the situation left to us by structuralist reflection. Its deliberately insulting ideological universalization has called forth an equally insulting "first wave" counter-movement. Mikel Dufrenne's plea "on behalf of man" (*Pour l'homme*) was published in opposition to the anti-human currents in contemporary philosophy stirred up by Foucault and Althusser. Another example is Raymond Aron's snide accusation that Althusser has only passed "from one Holy Family to another" (*D'une sainte famille à l'autre*) in his attempt to apply structuralist interpretation to Marx.

The problem put to "humanism" has been considered in a much more positive vein by the "personalist" followers of Emmanuel Mounier, the editors of *Esprit*, since 1963. The editorial group spent a whole year studying Lévi-Strauss' *The Savage Mind* (*La pensée sauvage*) and published their reflections together with their collective interview with the controversial figure himself in their November, 1963, issue. The popularity of this enterprise was so widespread that the entire issue was reprinted in 1968. In the meantime, the same group continued to consider structuralism, and published an issue entitled "Structuralisms: Ideology and Method" ("Structuralismes: idéologie et méthode") in May, 1967. In his introduction, the editor remarked:

> In November, 1963, before faddishness took over, *Esprit* devoted a special issue to structuralism which remains an essential introduction to the understanding of Claude Lévi-Strauss' thought. Since then, structuralism has developed, and the 'system' of Michel Foucault has been added to it. Thus, we devoted our Congress of December, 1966, to the various currents of thought provisionally allied in structuralism: the systematization of Michel Foucault, the epistemological Marxism of Louis Althusser, leaving the psychoanalysis of Jacques Lacan for later study. Our intention was not a counter-offensive at all costs; we wanted to examine these theses to question them. We believe that it is better to 'go through' them rather than to reject them. Their contribution to the human sciences is incontestable, and the reaction against an untrammeled affirmation of the self seems to us healthy. However, insofar as they pretend to reduce the meaning (of the subject) and the subject himself to their permutations, it is indispensable to proceed to clarification and to make appear the philosophy or philosophies which hide behind the billboard of methodological rigor....

One of the chief spokesmen against the universalization of formal, structural principles to cover the entirety of human existence has been Paul Ricœur, himself a member of the "team" of *Esprit*. His chief thesis has been his insistance precisely upon *la parole* as the link between purely formal structures (represented by the dictionary and the grammar) and the lived world of the speaker. By this thesis, he was taking into account the forgotten reality that while language, understood in terms of "social facts," "linguistic facts," and so on, can only be studied if we prescind from the fact that language and society do not change, they in fact must and do change; and that it is the *creative* use of language which each speaker achieves within the generality and constraint of the language system which operates the "drift" of language and creates its well-defined "slope" of change.

Between Ricoeur and Lévi-Strauss there seemed to be no means of compromise, in spite of the fact that Ricoeur's position was far from the extreme of Sartre's thesis of utter and unqualified creativity as over against all constraint of culture and society. It was a little-known philosopher, M. Van Esbroek, who recently showed the complementarity between the two seemingly irreconcilable viewpoints. Surprisingly enough, it was in the wake of the *theological* reflections of Henri de Lubac upon the meanings of the bible that Van Esbroek has been able to analyze the methodological presuppositions underlying both approaches.

Thus, a second, and hopefully more peaceful, generation of thinkers in psychoanalysis, ethnology, linguistics, and literary criticism (insofar as they make use of forms of structural analysis) is making itself known and appreciated. One thinks of the French genius Jacques Derrida when considering how best to sidestep creatively the acrimonious debates of the last ten years. One also thinks of François Wahl, the editor of *What is Structuralism? (Qu'est-ce que le Structuralisme?*) who has been able to demonstrate what is really sought by structuralist propagandists when one can get behind their deliberately offensive and clumsily shocking expression. He suggests that it may be possible to constitute true scientific enterprises in areas of the human sciences which have heretofore been recalcitrant to scientific definition.[16] Gritti and Toinet have recently underlined also the healthy aspects of the structuralist system (besides the incontestable value of structuralist method). They find, for example, that Lévi-Strauss' discovery of an unconscious rationalism in "savage" thought paralleled by an unconscious savagery in "rational" thought provides a healthy corrective to the pretension of Western Culture to be the apex of cultural development and civilization. Analogically, they find a certain healthy integration of humanity between the "sane" and the "mad" existing in the same universe in Foucault's *The Universe of Madness* (*L'univers de la folie*).

Van Esbroek, Gritti and Toinet, Wahl and others are all part of what Pierre Watté calls the "second wave" of structuralist discussion. Of this second generation, perhaps the most impressive figure is Piaget, whose

[16] Wahl and others refuse to define "structuralism" *a priori*, as we have done here for the sake of clarity. He remarks drily that we can count two positivist structuralisms (one accusing the other of empiricism), one frankly rationalist, two at least which proclaim a subversion of the subject (one accusing the other of reductionism), one philosophical structuralism (in the classical sense), and quite a few which make a claim to refute all philosophies. What is alone significant, in his view, is that all of these deal with sign-systems, the sole requisite being that they be reinterpretable in terms of language understood as a system of relationships of signifier and signified in a communication network. His own opinion of the "second generation" structuralisms is that they place emphasis upon the signifier rather than the signified, thus operating a whole series of revolutions in thought—the reversal of the classical notion of the priority of "that which is signified" over "that which signifies" being far from the least important.

book (entitled simply "Structuralism") has received the highest praise not merely for its timeliness and comprehensiveness, but especially because it reminds us that, as Jean Lacroix noted in his review, while talking about structuralism we have forgotten structure—a totality characterized by a multiplicity of internal and self-regulating transformations. Current lack of clarity in the debate about structuralism stems not so much from the historical or linear development of structuralist theories, as they have been traced in this chapter, as from the simultaneous diversity of cultural centers from which they emanate, especially linguistics, mathematics, and the history of philosophy. (These diverse originating centers of structuralist methodology have been identified by Gueroult, and Boudon as shown that together they are incapable of defining a *general* structuralist method.) The basic question, identified by Piaget, is whether a structure (if there is one to be found) is *given as such*, pre-formed, in such a manner that we can only describe it but cannot explain it. In showing masterfully that there cannot exist a "structure of all structures," Piaget has shown as well that "structures," in any case, are *not pre-formed*, but *constructed*. Lacroix sums up: "The true being of all structure is structuration." He concludes that the fears of the personalist are unfounded; "structures do not kill the subject... do not oppose operational genesis, but demand it." Radically, it is human intelligence, rooted in the human organism which is the prototype of all structuration, which operates the permanent construction of the world.

In conclusion, the structuralist model of man and his works is not yet finished. For a while, it seemed that the completely determined structure (universally found, universally applied) was going to win the day; then a sustained counter-attack from a phenomenological point of view forced a reexamination of the philosophical implications of the structuralist method. Since the philosophical importance of the questions raised by the structuralist enterprise have not yet, then, been adequately answered, work continues; and since, at the same time, there are not yet final answers, the polemic continues as well, side by side with the positive search for an answer to the question, "What can man today say about himself?"

SELECTED ADDITIONAL READINGS

FOR CHAPTER VI

Bloomfield, Leonard, *Language,* London, 1933.

de Saussure, Ferdinand, *A Course in General Linguistics,* trsl. Wade Baskin, New York, 1959.

Durkheim, Emile, *Rules of The Sociological Method,* 8th. ed., trsl. Sarah A. Solovay and John H. Mueller, ed. George E.G. Catlin, Chicago, 1938.

Goddard, David, "Conceptions of Structure in Lévi-Strauss and in British Anthropology," *Social Research,* 1965 (32), 408-427.

Hymes, Dell, "Why Linguistics Needs The Sociologist," *Social Research,* 1967 (34), 632-647.

Lantéri-Laura, Georges, "History and Structure in Anthropological Knowledge," *Social Research,* 1967 (34), 113-161.

Lévi-Strauss, Claude, *Totemism,* trsl. Rodney Needham, Boston, 1963.

—. *The Savage Mind,* Chicago, 1966.

—. *Structural Anthropology,* trsl. Claire Jacobson and B.G. Schoepf, New York, 1967.

—. *The Raw and the Cooked:* Vol. I: *Introduction to the Science of Mythology,* trsl. John and Doreen Weightman, New York, 1969.

—. *The Elementary Structures of Kinship,* trsl. J.H. Bell, J.R. von Sturmer, ed. Rodney Needham, 2nd ed., Boston, 1969.

Olshewsky, Thomas M., ed., *Problems in the Philosophy of Language,* New York, 1969.

Ricoeur, Paul, "New Developments in Phenomenology in France: The Phenomenology of Language," *Social Research,* 1967 (34), 1-30.

Rorty, Richard, ed., *The Linguistic Turn: Recent Essays in Philosophical Method,* Chicago, 1967.

Sayre, Kenneth, *Consciousness: A Philosophic Study of Minds and Machines,* New York, 1969.

Whorf, Benjamin Lee, *Language, Thought, and Reality,* ed. John B. Carroll, Cambridge, Mass., 1964.

Yale French Studies, Nos. 36-37, 1966: "Structuralism" (with extensive bibliography).

Ysseling, Samuel, "Structuralism and Psychoanalysis in the work of Jacques Lacan," *International Philosophical Quarterly,* 1970 no. 1 (10), 102-117.

CHAPTER VII

THE PRESENT STATUS OF PHILOSOPHICAL ANTHROPOLOGY

A PROLEGOMENON

It is difficult to write on a subject which hardly exists, except in the spirit of its practitioners. In the English-speaking world there are no Chairs of Philosophical Anthropology, and courses with the title are rare. The philosopher, it seems, has some reason to expect the accusation of poaching in the fields of the "true" anthropologist, or psychologist, or economist, or what-have-you, since he shares the data of their sciences with them.

The notion of a philosophical anthropology did not spring into existence suddenly, without antecedents. The subject is an outgrowth of what used to be called, in some circles, "philosophical psychology," and more recently, "the philosophy of man." Philosophical psychology was an outgrowth of the Scholastic enterprise of "rational psychology," a manual treatment of the peculiarly epistemological and psychological works of Thomas Aquinas (for example, the treatise *de Veritate*, or the commentaries upon Aristotle's *de Anima*). In the course of time, the manuals were retouched, reorganized or remodeled, keeping the basic Aristotelian and Thomistic orientation, together, often, with the Kantian-Thomistic synthesis of Maréchal. Lonergan, Coreth, Rahner, Lotz and Donceel, among others, have been the guiding lights of this movement in modern times; but the movement, with all its accretions, its growing respect for "existentialism" and "phenomenology," its increasing abandonment of antequated terminology, still searches for "what makes man properly man," perhaps for an "essence." But in the context of the "*philosophia perennis,*" in which essence is constituted through genus and specific difference, the old definition of man as "rational animal" no longer suffices, and the effort at explicitating the definition with the help of modern scientific experimentation succeeds only in demonstrating the inadequacy of the original definition. For example, Joseph Donceel's book, *Philosophical Anthropology* (the only book in English, to my knowledge, to use this daring title), in spite of its effort at updating, still shows its functional, "*agere sequitur esse*" background and the recursive logic of "*ab esse ad posse*" in filigree: "the senses," "the mind," "the personality," knowledge," and so forth. This topology perhaps, saves the effort from complete electicism; but the "perennial philosophy" which was to have been its inspiration has left only the sedimentation of chapter headings. Is man, in this view, just the *sum* of these functions? If so, what is the principle of summarization? If not, from what hidden preexisting and pre-known totality do these functions issue? We have had occasion to remark upon the *de facto* presuppositions of this kind of enterprise.

Etymologically, of course, "anthropology" has meant the science about man, the study of man. However, many sciences study man, and many of them make a claim of exclusiveness, in varying degrees. A *philosophical anthropology,* properly speaking, must first of all exercise a degree of critical reflection upon the truth-claims of the various sciences which tend to universalize the results of particular researches into an explanation of Man, "the total man." The models which issue from the particular sciences of man serve here as the subjects of the preceding chapters, and critical discussion of them was the task of the philosophical reflection attempted.

It is not to be imagined, however, that critical reflection upon received models pretending to exclusiveness and universality is the sole purpose of the philosophical undertaking here. A science which calls all other sciences into question is one which makes a pretence of ascendency over them all. To accept this mission without revealing the sources of one's own critical stance is to accept only half of the philosophical endeavor. Consequently, a second effort of the present volume is to put into the foreground not a negative reflection upon the results of particular sciences or theories, but a positive effort to understand their foundations as part of the critical undertaking itself, and hence of the basic intention of the studies concerning man.

The purpose of the present volume is somewhat different from its likes, and the difference warrants a different book. It is too easily said that "many sciences study man." What is often overlooked is that the many sciences which study man (let us say, his psychology, his social character, his unconscious, his origins and physical and cultural development, his ecology, and so on) furnish complete models of who, or what, man is. *They can pretend, as sciences, to nothing less* than completeness and universality; any knowledge that we might have of "man" outside of the knowledge that we have of him in any of these sciences is simply "unscientific" from the point of view of the science involved. The option of the present book is that the final explanation of "man" lies outside of *all* the possible scientific views of him because it lies within the *origins* of any and all the sciences, including the science of philosophy. The currently fashionable models of man appear to some to be reductivist (the "nothing but" type of explanation); but, in fact, we cannot expect them to be anything else. Such models, through their own coherence and the coherence and rigor of the methods which serve in their elaboration, make a claim to ultimacy, and we had to take them seriously. It is difficult to impress the importance of this view upon beginners in philosophy; we owe to Edmund Husserl, in his operation of the "transcendental reduction," and his "bracketing of the world of the *natural attitude,*" the sole possibility we have of bringing to consciousness the realization that we possess a non-scientific pre-knowledge of our world and of ourselves which we are explicitating through our objective scientific endeavors. Without the transcendental focus, such consciousness would simply "go without saying," and we would never have explicit knowledge that there is the possibility of *reflection,* and consequently, the "objective view" of the universe which we have scientifically would be the only one possible.

There is, then, a series of questions which must be answered here, or at least a series of problems which must be clearly distinguished. The first question is one of perspective. It was said above that "many sciences study man." Traditionally speaking, each of these sciences chooses a perspective which constitutes the "formal object" of the science in question. There is thus one clue provided for the success of the enterprise of philosophical anthropology: there is only *one* object, properly speaking, the "*material object,*" man himself. But the initial problem also arises here; a material object cannot be studied "in itself," but only under the modality of some formal perspective. Consequently, to carry out the implications of the task, it seems that one must add that this one object should be treated from all possible points of view. True enough, and that will make the enterprise a truly interdisciplinary one. However, one must be careful not to reintroduce incoherence through eclecticism, that is, by adopting simultaneously incompatible or contradictory points of view. This, in my opinion, is the heart of the problem, and the precise fault of the "human sciences," taken as an agglomerate, today. Though it may be true that an "object" (of study) is the sum total, or total structure, of all possible perspectives upon it, Jérome Grynpas, in a closely reasoned passage concludes that the project is impossible:

......it comes down to saying that there would be a list of various domains relative to 'man,' and that the sum of the knowledge relative to each of them would lead me to the discovery of what one must know about 'man.' The one who would be called upon to speak would then present himself as the Super-Competent, the one capable of assimilating all the subordinate competences.... How can I know if such an operation is possible, or even if it makes any sense? [1]

Theoretically, of course, such an operation might be considered possible, especially if one invokes a transcendental ego as the operator; but if one looks a bit more closely at the problem (following Husserl), one discovers a situation as follows: A real thing cannot be presented "as it is in itself" to consciousness or perception except in a series of *successive* presentations, that is, in a multiplicity of perspectives, none of which present the thing "just at it is in itself," but only "within the limits within which it presents itself," that is, in one perspective at a time. Of course, as Aron Gurwitsch points out, "These perceptions enter into a synthesis of identification with one another, and it is by, and in, this synthesis and the parallel synthesis among the corresponding noemata, that what appears successively constitutes itself, for consciousness, into this real thing which it is, one and identical as opposed to the multiple perceptions and also the multiple noemata." [2] Thus, by analogy, we may represent "perceptions" as the perspectives of the various sciences upon "man," the *noema* of each corresponding to a specific kind of *noetic* activity which constitutes the science, and the multiple perspectives upon "man" provided by the sciences would, hopefully, coalesce, by a series of mutual identifications, into a constituted appearance in consciousness of "this real thing which it is."

Practically, however, a problem appears in operating the "parallel synthesis" on the *noetic* side. While the real thing which is man may become constituted in consciousness, under the guise of a synthesis of its multiple noematic presentations, all conscious noetic perspectives are impossible *simultaneously;* the identity of noemata as constituting presentations of real things, "identical as opposed to the multiple perceptions and also to the multiple noemata," still leaves the noetic acts multiple. The mystery of how we know "man" as a totality while at

[1] Jérome Grynpas, *La Philosophie; sa vocation créative, sa position devant la science, ses rapports avec l'homme et la société d'aujourd'hui,* Paris, 1969, 130.

[2] Aron Gurwitsch, "On the Intentionality of Consciousness," *Phenomenology: The Philosophy of Edmund Husserl and Its Interpretation,* ed. Joseph Kockelmans, New York, 1967, 8.

the same time knowing that each separate scientific perspective is *not* a total perspective cannot be solved scientifically. The basic reason for the impossibility of scientific resolution of this question is that, by its options, each science accepts the task of restricting itself to one single noema as its object of investigation; this constitutes the "formal object" of the science and it can have no other and still be true to the scientific enterprise. In fact, one of the efforts of this book has been to indicate that, by and large, the sciences of man have not been entirely faithful to the truly scientific project.

Space does not permit further discussion of the impossible simultaneity of noetic accessions to "man as such" or the "total man" in the scientific context; it is important, however, to point out at least one primordial incompatibility: that between man as "object," the formal object of the various sciences of man, and man as "subject," the actor and instigator of these same sciences. Insofar as man has indeed become an "object" of science he is out of the game, or at least is no longer worth bothering about. Let us consider an example. The experimental object, called the "subject" in an experimental situation in psychology, is not really capable of providing a fruitful point of departure for philosophical reflection, and there is grave doubt as to whether he is even useful as a source of scientific reflection. In point of fact, he does not really exist. He is, and is considered both by psychologists and by philosophers, simply the point of reference for an experiment, the anonymous source of some reaction purposely elicited from him and capable of quantified expression. To explain: a "response" is provoked from the experimental subject; the response is the "given" of the experiment, the "object" of study. Both psychologists and philosophers accept as a matter of fact the objectification of the subject of positive science in general, and of experimental psychology in particular, although they by no means take the same attitude toward the state of affairs thus accepted. For example, De Waelhens says:

> Scientific knowledge is essentially the theory and the system of the objective dimension of reality which, in its full sense, is first of all and primarily encounter. It is possible, in fact, to distinguish abstractly, in the midst of perceptive coexistence, an objective face, the dimension of the other. Its explicitation is the work of science; and that is why it is constitutive of all 'positive' knowledge that it considers that which it studies as simply a given.[3]

[3] Alphonse De Waelhens, *Existence et signification*, Louvain, 1958, 108-109.

De Waelhens explains, following Husserl, that what happens to the "given" is this: "Explanatory procedures which take place with regard to the 'given' constitute an explanatory system elaborated without explicit consideration (along with the given itself) of the act which refers to or begets the given."[4] He deftly summarizes the significance of positive science insofar as it is explanatory: "It abstracts absolutely from the intentional movement which carries us toward things, and leaves aside the effort which makes them or lets them appear. In other words, it considers the object as simply given and ignores the fact that is reveals itself in the midst of an encounter or a coexistence." [5]

The inaccessibility of the subject-object (or object-subject) in experimental psychology is deplored by the philosopher, but it was simply taken for granted by the late psychologist, the famous Albert Michotte, who felt constrained to apologize for mentioning a fact so banal. The "sensory content" of the perception-event, he reminded the 15th International Congress of Psychology, "does not exist except for the one who 'lives' it; and consequently, for everyone else . . . it is an unknown, of which only certain indices can make us suspect the presence." [6] The most efficient of these indices is, of course, language, but the words by which the subject reports his experiences can be accepted only as "differential reactions which inform us of the presence or absence of a certain phenomenon X or of a certain aspect of phenomenon X; and we do not attempt to imagine or represent what that can be for the subject." [7] The subject, then, is hidden *as subject* from the experimenter, who can have no idea of "hidden life" of his subject or any apprehension of what his world means to him. In Michotte's experiments, then, the subject as such does not exist. He must disappear, for his role is only a point of reference for the experiment. He does not count in the experimental situation. His "differential reaction" is the only reality to be considered. His response is tabulated with that of others and presented as part of the statistical report of the perception-event. But he is out of the game.

In performing this type of experiment, the experimenter must assume that the subject can be detached from his responses, that the subject's actions (the facts, the given) are somehow distinct from him. The experimenter must further assume that no one can know what a given experience

[4] *Ibid.*, 105.
[5] *Ibid.*, 107.
[6] Albert Michotte, "Réflexions sur le rôle du langage dans l'analyse des organisations perceptives," *Actes du XVème Congrès International de Psychologie, Bruxelles, 1957,* Amsterdam, 1959, 17.
[7] *Ibid.*

is for the subject, except the subject himself; and the subject cannot communicate with the experimenter, for his word is not *logos*. In his study of the illusory interiority of the "subject" in experimental psychology, Georges Thinès remarks that the abandonment of the subject coincides with the doubling of the subject in scientific thought. That is, there exists, side-by-side, as it were, with this mysterious and inaccessible person, the "subject of the experiment," a docile intermediary who "reacts," who "as soon as he begins to respond, already is repeating without knowing it that which the science which uses him rigorously dictates." In other words, a programmed responder in the skin of the ineffable person. Because of this doubling, language loses its "apophantic" character, and with this loss man himself disappears from scientific consideration.[8]

These remarks seem valid enough as applied to the experimental subject, insofar as he is considered the mechanism which triggers a response to be measured *quantitatively*. Are they valid as well for the "qualitative" subject? The qualitative subject is still the object of an experiment, minus a functional variable; as such he is still liable to the objection registered above: that in experimental psychology, the subject is out of the game. The paralogism of "individual psychology" is that it must become differential psychology; and the fact that the statistically significant element in the experiment is no longer the response of the individual simply, but the differences among the responses of individuals, does not change the state of affairs. Here it is the differences among subjects which are considered without reference to the subjects themselves. Consequently, the first question, that of perspective, must somehow be resolved if we are to proceed. Not only does the perspective of "observable differences" not coincide with that of "individual meaning" in the scientific study of man. This is only a symptom of a deeper problem. More basically, we must decide what perspective shall be most adequate for the study of man. It is at once clear that the role of philosophy as *reflecting upon the data of science* in order to answer the question about man is no longer a supportable option. The results of the scientific endeavor are already prescribed by the originating options of scientists.

Is it, then, possible, as an alternative problematic, to lay the groundwork for a philosophy of man which can, in turn, serve as a groundwork for the sciences of man by uncovering the vectors determining the

[8] Georges Thinès, *La problématique de la psychologie,* The Hague, 1968, 116-118. Thinès' reflections here are based on Heidegger's *Einführung in die Metaphysik,* in which analysis reveals language as the utterance of Being, and man as the utterer.

horizon within which the data are accessible? [9] Such a project is indeed possible, since, in any case, philosophical presuppositions (unexplicitated evidences) lie at the origins of all sciences. But the assertion of this possibility carries a *proviso:* Provided philosophy itself be included in the class of sciences obliged to clarify their own presuppositions and to justify them. If philosophers can be clear about their own presuppositions, and justify them, they shall have a coherent basis upon which to launch a critical study of the various claims to define man. In what follows I shall attempt to sketch out the basic presuppositions of such a philosophy of man which, in its very generality, may hopefully serve as a unifying "point of view" which informs all the perspectives upon man without compromising the methodologies and formal aspects proper to each one.

Before discussing the philosophical presuppositions which I think lie at the basis of any viable philosophical endeavor in the study of man, a few introductory remarks of a historical nature may serve to focus the meaning of the notion of philosophical presuppositions as the unifying principles of the human sciences.

Although, philosophically speaking, the nineteenth century was monistic in practically all its manifestations, almost all the intelectual movements at the end of the century, on the contrary, were pluralist. This is the assessment of Bochenski. If the nineteenth century was especially materialistic in outlook, as well as monist, the crisis of 1900 announced a solidly established interest in what Bochenski calls "spiritualistic personalism." This was an especially European phenomenon. "Spiritualistic personalism" in America has remained largely underground since the turn of the century. Its spokesmen here were predominantly ecclesiastical, with theology, not philosophy, their base. But even in the United States, as the twentieth century began, philosophy was at a standstill before the scientific revolution of the preceding century. Under the pressure of positivism, philosophy then took a much more modest stance than it had during its own three centuries of power. The rationalism of the preceding century, for the most part, was abandoned. Philosophers were reduced to teaching simple inventories of earlier systems, or joined the positivist enterprise and turned to analysis and logic.

Thus, the "sciences," including psychology, became liberated from

[9] This is a very shorthand way of stating the whole thesis of Johannes Linschoten's book, *On the Way Toward a Phenomenological Psychology: The Psychology of William James,* ed. Amedeo Giorgi, Pittsburgh, 1968. The programmatic thesis here is that "access to phenomena lies in experience, the basis of the phenomena lies in the body," behind which is Husserl's essay on "The Body as Center of Orientation," in his *Ideen II.*

the domination of systematic philosophy; and for the next half century or more "scientific" studies of man held first place, while a kind of "philosophical anthropology" inspired by James and Bergson went its own hidden way.

At the present time, the "life sciences," as they are now called, have become increasingly powerful, as the project of "human engineering" increasingly attracts the technician (propelled towards it, likewise, by public demand). At the same time, practitioners of the "life sciences" have become enmeshed with their own methodological limitations, especially the rejection of "subjectivity" and the consequent loss of a viable concept of "life" itself. On the contrary, philosophy has been methodologically regenerated through phenomenological method. Presently, instead of a turn of the century pluralism, two points of view regarding man are at present in dialogue, two radically different and complete world views, one philosophical, one scientific, are being transformed, each by the other, through a dialectic in which philosophy and the human sciences—while remaining indentifiably distinct—are nevertheless assimilating one another methodologically.

One symptom of the present situation may be found in two major review articles within four recent months published in the *New York Times Book Review* deploring the contemporary reversal of faith in the truly "scientific," dispassionate, objective, detached character of both the social sciences and even more the natural sciences (especially nuclear physics, microbiology, and biochemistry). The diagnostic intention of the present volume has been in part to argue against the reality of the supposed "dispassionateness" of the kind of objectivity we have been led to believe is actually practiced in the social and natural sciences, and in part to suggest that the cries of alarm against the dangers of "subjectivism" are unfounded in that they issue from an impoverished notion of subjectivity—which is confused with "subjectivism."

The precipitating factors in the present disarray of the "life sciences" are likewise historical. For example, psychology, a primary example of the "scientific," i.e. "objective," view of man seems to have reached the extreme of its methodological limitations, with even the hardest-nosed practitioners (for example, successive presidents of the APA) increasingly alarmed at what seems a man-made monster—their own science—gone out of control. (There are, of course, striking exceptions.) How did this happen? As we have seen, with Wundt, "general" psychology, as a part of the philosophical enterprise inherited from the Middle Ages, gave way to "experimental" psychology. McLeod, in his now famous article on "The Phenomenological Approach to Social Psychology," characterized the effort in its origins as already limited by interrelated classical biases:

......the belief that something small is more fundamental than something large —the atomistic-reductive bias; (2) the belief that a differential variable in the physical world or a differentiable receptor in the organism implies the existence of a correspondingly differentiated unit of experience—the stimulus-receptor bias; (3) the belief that that which is genetically early is more fundamental than that which is genetically late—the genetic bias.

According to McLeod, the "atomistic-reductive" bias was assimilated directly from nineteenth century physical science. The physicists' search for physical "elements" was transformed in the psychologist's laboratory into a search for the "elements" of mind and perception. The "stimulus-receptor" bias was partly the by-product of the laboratory method itself, and partly the consequence of the Aristotelian theory of sensation, namely, that there is a correlation between a stimulus variable and an elementary sensory reaction, with the corresponding theoretical conclusion that "elementary" experiences are due to the activation of simple reactor mechanisms (together with the conviction that there *are* such elementary experiences, classically termed "brute sensations"). The more tenacious "genetic" bias was derived from the elaboration of evolutionary theories, but was even more deeply linked with the much older search for "origins," or "first-time experiences," as the ideal of science and the "naive subject" as the ideal game for the scientist: the conviction that earlier forms of phenomena are somehow "truer." In consequence, the psychologist became convinced that everyday perception is tainted with past experience, that immediacy is found only in first-time occurrences; the psychologist felt justified in his unwillingness to deal with "everyday meanings." This tendency became even more marked in the practice of psychotherapy, especially psychoanalysis, where everyday meanings are assumed to be distortions of primordial experiencings and are to be transformed into the language of an original myth in terms of which primordial experiencings are supposed to be lived.

But within the new "scientific" view of man, with its classical presuppositions, there were tensions. Just as the earlier Empiricism of Locke, Hume, and Mill was set over against the Intuitionism of Leibniz, Kant, and Fichte, so the later Empirical Geneticism of Helmholz and Wundt was set over against the intuitionist-like Nativism of Müller and Stumpf. (The comparison is Boring's.) Dewey's Functionalism came as the first revolt against Wundt's psychology, but was submerged by the noisy Behaviorism of Watson and an ultimate takeover by Skinner's Operationalism. Only in the phenomenology of Merleau-Ponty (*The Structure of Behavior*) does the earlier revolt embodied in Dewey's 1896 article, "The Reflex Arc Concept in Psychology," reappear as a worthy contender for attention, buttressed with a long line of equally impressive experiments and reflections demonstrating the self-contradictioness of all experimental models.

Another example: The "social behaviorists," Mead and Parsons, following the same post-Wundtian road as Watson, achieved an impressively systematic, modern, "scientific" sociology. But not until Winter's successful effort to escape from their captivity by the relativization of "science" with the help of the "social phenomenologist" Alfred Schutz (the pivotal figure in the attempt to correlate observation with understanding) did sociology realize that it had somehow lost both man and society somewhere along the line, probably in the matrices of the correlation box.

Another branch of the same science, meanwhile, following the establishment of the "social fact" by Durkheim, after its structural pilgrimage along the same Watsonian road through linguistics, anthropology and psychoanalysis, not only does not seem to care whether both man and society have somehow been misplaced, but concludes, with Foucault, that neither "he" nor "it" was ever there in the first place—or if they once were, they either are no more or are on the way out.

The point of these reflections is that some critical evaluation of method and purposes in the human sciences was absolutely essential, together with some unifying point of departure and some unifying perspective. The first of these tasks cannot be undertaken on the basis of the results of the sciences in question, it has been remarked, since such results are already predetermined by the formality of the object of the sciences and of the observational methods employed in their elaboration. The latter task, it seems, is feasible; but evaluation and unification must have a philosophical basis. None of the separate sciences which deal in any way with "man" and his projects can provide such a basis, even though most of them pretend, with equal universality, to do so. Yet at the same time the philosophy in question, since it is to be a philosophy undergirding the human sciences, must in itself be a philosophy of man. Moreover, such a philosophy cannot be simply a system of ground rules dictating the possible directions and methodologies of research, on the model of earlier versions of systematic philosophies. It must provide a critique not only of the presupposiitons (understand here: unreflected or forgotten evidences) of science, but of itself as well. If it turns out that the philosophy of man, even when purged of its dogmatisms, cannot be presuppositionless, yet philosophy is best equipped to achieve some clarity about its own presuppositions. In the last analysis, perhaps "option" is the best word to use when choosing a perspective for philosophy, and "attitude" is the best term for its method.

The option of the paragraphs to follow is that a philosophy of man, a "philosophical anthropology," is the only philosophy relevant today. Philosophies can continue, of course, to speak their arcane languages and construct their own abstract categories; but can they speak meaningfully to anyone? Are the constructors of such philosophies not liable to the same ridicule which Kierkegaard directed toward the lovers of "The System" of his day, that they constructed magnificent palaces of thought yet lived in kennels nearby? In every human dialogue there is a "hidden agenda" and that is the question, "what is in it for me?" This is as true for the scientific dialogue (does anyone live in "the world of science"?) as for the philosophical dialogue (are philosophies *of* life more important than philosophies *about life*?). Can we any longer continue to luxuriate in the purely aesthetic contemplation of the magnificance of systematic thought? It is here urged that philosophers must think in the contexts of other sciences, especially, in the present instance, in the context of the human or life sciences. In taking this option, we are going resolutely against the current of positivist influence in American philosophy, with its technological attraction for formal logic and analysis. We are considering it, for our purposes, an unfruitful enterprise. I should like, on the contrary, to assert that *all* sciences are human sciences, in the sense best explicitated phenomenologically (for example, by contemporary German philosophers like Gerd Brand, Hans-Georg Gadamer and Elizabeth Ströker, among others of other nations). Such authors resolutely abstain from considering "Science" or "Philosophy" as something "out there," and consider these as enterprises of human persons, undertaken with human means to achieve human ends. As a result of this option, science and philosophy are interpretable primordially in terms of the human intentions which they reveal. In the light of this option, one can maintain coherently that empirical science of whatever kind and philosophy are complementary. Not that they are complementary in the sense that there are distinct enterprises (or any number of sciences and philosophies) where each brings to light something which the other cannot attain. This is, in part, true; and Strasser has well shown this kind of complementarity in his *Phenomenology and the Human Sciences;* but the real answer lies at a deeper level. It is not enough that the philosopher know, for example, a little psychology and that the psychologist know a little philosophy so that each may in empathic dialogue with the other, at the boundaries where their individual enterprises touch, enrich his own and the other's thought about the meaning of man. It is impossible that a science already complete and autonomous in its own right be "completed" by another. (The problem is, of course, to be absolutely positive that the science in question is really complete and autonomous; the answer cannot, in principle, be provided by the science itself.) Nor, on the other hand, is it possible, after the painful liberation of the human sciences from philosophy that they fall once again into the all-encompassing embrace of philosophical dogmas, or that philosophy, after its liberation (through Brentano and Husserl especially) from mechanistic presuppositions should fall again under the spell of physicalist thinking. What founds both the human sciences and philosophy—and I maintain, all approaches to the meaning of man—must be found in something which gives them all meaning. The special problem, then, is not to make one science of philosophy and the "sciences" about man, but to found all of them in one primordial intentionality which makes sense of all of them while leaving them autonomous.[10] This is to say that, coupled with a primordial *option* in this book, there is a primordial *attitude* which is phenomenological. It is to say that "philosophy" here is understood not as system but as method, and furthermore, a method which seeks to explicitate the foundations of sciences and philosophies.

The only claims which philosophy can make to leadership in the total enterprise of understanding man is its capacity to explicitate its own presuppositions. Of course, a contrary view is maintained even by scientists, whose experience of *Grundlagenkrisen* ("basic crises") in their fields always provides the occasion for a scientific reformulation of the fundamental principles of their scientific disciplines, and thus leads them to contest any philosophical pretension to meddling with them. However, we are indebted to Husserl for the insight that it is precisely because science reformulates its own foundations that it obscures its own view of what is really fundamental. (That is, the foundation of science is always *relative to the science* in question, and not to any ultimate or absolute foundation; the latter is found only through an investigation of the ultimate sources of all knowledge.) Thus, if the necessary pre-

[10] The concept of "intentionality" is perhaps best explicitated for the beginner by Aron Gurwitsch, **"On the Intentionality of Consciousness,"** in **Marvin Farber (ed.),** ***Philosophical Essays in Memory of Edmund Husserl*** (Cambridge, Mass.), 1940, 65-83. The article has been reprinted in Joseph Kockelmans (ed.), *Phenomenology: The Philosophy of Edmund Husserl and its Interpretation.* Garden City, New York, 1967, 118-137.

suppositions of a philosophy of man can be clarified and justified, its claim to be basic can be validated; and this is why "philosophy" in this book is understood as being based on "option" and why its character is understood as being methodological, or an "attitude."

The presuppositions of any philosophy, I maintain, involve a fundamental attitude toward myself, the other, and the world. The most fundamental evidence here is that *the universe is not a thing, nor even a system of things, not an object or a system of objects, but primordially an interpersonal world, a world of and for persons.* The problem of these persons is that they have constituted a world of objects and then have forgotten the act of constitution. In our world of technological objectivity, the human person, as the originator of objectivity, has become confused with his own creation. Our inability to remember our act of creation is coterminous with our inability to remember having forgotten. The main point of the following paragraphs, then, is an effort to remember, perhaps by negation more than by affirmation, what are the ineffaceable bench marks of our passage through the world in which, and through which, persons are in communication.

Regarding "myself:" The insoluble problem of Solipsism which so engaged neophytes in philosophy at the beginning of this century has, fortunately, been abandoned. It is understood that no one can "prove" his own existence; and unless one's own existence is presupposed in the very beginning of philosophy, no progress can be made. But the very abandonment of Solipsism as a real problem has brought to the fore a Cartesian emphasis upon the *Ego* as originator of the totality of the significance of the personal and natural world. Now, however, it is clear that "I" (the "I" of the "*Cogito*") cannot be the starting point in philosophical investigation. Both the presuppositionless beginning and the absolute beginning implied in "my" being the starting-point of philosophy are impossible. The absolute beginning involves the question, "Can I know anything?" which is always raised when philosophical systems in vogue, or the "inherited conglomerate," or even the more lucid statement of a society's "desirables" have exhausted their possibilities and given way to a general spirit of skepticism or anguish over the breakdown of received models and definitions. The response to the question, "Can I know anything?" consists in returning to what is thought to be the least contestable minimum of affirmation as a starting-point in philosophy, and implies one or other variation of the *cogito* argument. (St. Augustine used it even before Descartes.) Heidegger, in our time, has traced the history of the failure of the absolute beginning of a philosophy which asserts that the primordial evidence in human knowledge is "I think," and the illusory conclusion, "therefore I am." "I am" is precisely not an epistemological statement nor a logical conclusion, for it is the very presupposition of one's thinking. The primordial reference to "I am" as an ontological statement rather than to "I think" as an epistemological statement is, then, a first step.

A second step is to ask where, precisely, I am. Heidegger's apparently simple answer is that I am simply "there;" and this is the beginning of his ontological analysis of how it is that I am there, and what is the mode of my being there. We shall have to leave to the reader the task of following the steps in Heidegger's analysis of how Western thought has foundered in the shallows of its own egoistic launching site. The conclusion to be drawn here is more modest. The impossibility of an absolute beginning entails the impossibility of a presuppositionless beginning, since we can at least recognize the primordial reference to *being* in all knowledge and affirmation. Philosophy can then be defined as a reflection upon the pre-reflexive, pre-philosophical, pre-scientific experiencing of being, that is, upon *experiencing* before any kind of conscious thematization. If philosophy is a radical and transcendental thinking, that is, a thinking upon the *a priori* conditions of possibility of all thinking and all experiencing, then the experiencing which is reflected upon must be experiencing in the largest sense. It is the experiencing of my insertion in being—concretely, the experiencing of myself and the other in the world. "Experiencing," then can be discussed in its ontic or personal dimension, as that which founds *for us* all truth and all certitude. On this level, the "experiencing" of human persons or collectivities as thematized in various ways by the human sciences is found meaningful in its own way (the given is studied in its givenness within the limits in which it gives itself); but we need not, then, overlook or ignore the basic and metaphysical horizon within which the meaningfulness of our ontic experiencing is grounded.

"I am," then, grounded in being before "I think" in any way; the impossibility of an absolute beginning in philosophy is explicitated. However, the "I am" which has operated the overthrow of the "I think" tradition cannot be the absolute starting point in philosophy either. This would lead back to the Solipsist problem. Here it might be useful to plant a few guideposts to show how the way leads out from that problem. This step, then, is an attempt to undo the harmful effects in philosophy of the "idealist bias" (to use Dondeyne's terminology) which has always regarded the opacity and the resistance-to-me of "the other" as a limit. The result of this attempt is two-fold. First, it makes unnecessary the affirmation of the other as the *term* of some reasoning (or other intellectual or empathic) process. In other words, there is no way to justify the presence-to-me of the other; and if he is not present as a "given" in the starting-point of philosophy, he tends to recur as a problem. The significance of this problem is easily overlooked in systematic philosophy; there, the other as other is simply a recurring problem. The unjustified and unjustifiable presence of the other in the system can thus be, unfortunately, overlooked, and "I-other" relationships treated in the same fashion as "I-allon" relationships. The second effect of rejecting the idealist bias, in which the other is simply a limit for me, is to force reflection upon the *radical* nature of my powers of affirmation and negation, that is, of treating affirmation and negation as modes of being rather than as logical categories. In the tradition which has continued the idealist bias, at least from Fichte to Sartre, the paradoxical nature of transcendence-immanence, positivity-negativity, and so on, has been accepted as a simple and unresolvable paradox, the irreducible index of human finitude and contingency, the utter surd. Beyond this point phenomenological description has hesitated to go. Traditional metaphysics has taken this situation as the radical starting-point for reflection upon the Absolute. For Sartre and Merleau-Ponty, on the contrary, the experience of human finiteness leads rather to the negation of God; for the only certitude available to the finite creature is his own freedom, and if man is free, God cannot exist. However, if the radical negativity of myself with regard to the other can be overcome and replaced by an even more radical analysis of affirmation and openness (concretely, if phenomenological description can identify those power centers of affirmation, of truth and freedom, of welcoming and creating which make man a historical being, as Dondeyne suggests), the problem is capable of solution. Philosophy as the explicitation of pre-philosophical experiencing will be viewed, then, as the thematization, both ontic and, *a fortiori,* metaphysical.

Granting that "I am" is a more basic affirmation than "I think," that is, granting that the real question of the *cogito* is a metaphysical rather than a simple epistemological one, and that the "I" who am "there" am "there" in all possible human dimensions, is it not possible to go beyond even Heidegger's correction of the tradition of Western philosophy and to affirm that "we are" is an even more basic affirmation than "I am?" I do not see how this can be denied; otherwise, as noted above, "you" become an insoluble problem in philosophy. If "we are" is a basic starting point, if without "you" there is even no "me," then the constitutive role of the other demands attention.[11] The "*Mitsein*" of the early Heidegger, where the other is found with me at work in the world, can serve to some extent; but it is far from answering the full dimension of the meaningfulness of the presence of the other with me. Here, then, we must leave, at least temporarily, the metaphysical consideration for the more proximate ontic one; there are more data on the level of pre-philosophical experiencing which can lead to the reflexive affirmation which has been made here.

It is at this point that a discussion of the "human sciences" in a phenomenological context is in order, in which the constitutive role of the other must be investigated. In general, one might well adopt the viewpoint that since "I am" (becoming a person) because "you are," meditation on the ontic experiencing of selfhood is appropriate, both in society (in terms of language, culture, history, and so on) and in one-to-one relationships, beginning with the earliest dyad and triad, and continuing a phenomenological description of the emerging person not as a myth or a function of language or society, but as a total organism, together with the possible dynamics of his possible disintegration and possible reintegration.

A deployment of the resources of the human sciences can provide a more lucid return to an ontological level of reflection, inasmuch as reflection upon the pre-reflexive is the task of philosophy, that is, inasmuch as human experiencing is the access to philosophy as it is to science. The "models of man" discussed in these chapters are a few examples of scientific efforts to expliciate how man is "there." In every case, beyond the *formulation* of a particular human "mode of being-in-the-world" there is the experienced reality of human being in the world which may be more or less adequately revealed in the scientific formulation of the experiencing. In every case, also, only one formal aspect of being-in-the-world can properly be considered ("the given . . . only within the

[11] The heretofore insoluble problem of the givenness of "the other" has been much enlightened, if not definitively solved, by Stephan Strasser's yet unpublished paper, "Feeling as the Basis of Knowing and Recognizing the Other as an Ego." Merleau-Ponty had partially solved the problem put by Mill, Avenarius, Lipps, Husserl, Malcolm and Ayer, but went only as far as recognizing "the other" as "body-subject," but by no means as "another ego." Thus, according to Strasser, even Merleau-Ponty remains in a "dualistic rut." Strasser's solution, based upon a conviction that human existence has an "underlying elementary structure which is not based on objectifying knowledge and discriminatory volition," results in a theory which contradicts the ordinary affirmation of the primacy of "selfhood"—of a primordial "self" which must somehow get in touch with "the other"—and asserts cogently that "the feeling subject feels himself by the fact that he feels his fellow-subject" first. We can only hope for an early public appearance of Strasser's paper so that his reasoning may be available to all.

limits which it gives itself . . ."); otherwise the enterprise is simply not "scientific" in any real sense. On the other hand, and at the same time, the unexplicitated *extra-scientific* knowledge we have of ourselves and others is constantly at work in the enterprise of interpreting scientific facts. In considering some of these scientifically elaborated models, we have made some attempt to indicate this inescapable duality in the heart of the scientific study of man, and to explicitate the unavowed and divergent options which govern it.

One might have proceeded otherwise. An equally respectable starting-point might have been a coherent Theory of Personality which takes into account the emergent reality of the person as the manifestation in consciousness of *a being* (the emerging self-consciousness of an individual as well as the emergence of the selfhood of the other in the consciousness of the scientific observer). One might then have passed on, without atomization, to a consideration of the emergence of societies, cultures, and civilizations on the same model. This would make possible a transition, without atomization, to a reflexive metaphysical consideration of the manifestation of *Being* (the ontic, or existential, manifestation or appearance of a person, a society, a civilization, as concrete, ontological, instances of the truth of Being).

However, this method would have suggested that "metaphysics" does in fact "*come after*" physics (in a sense more basically procedural than the place of "first philosophy" in the catalogue of the Aristotelian *Corpus*), that metaphysical reflection must, in a sense, wait until the data are all in. But we cannot wait; first, because in spite of the eschatological hopes of our empiricist friends, all the results will never be in—precisely because every established "fact" enriches and changes our knowledge of the world as it modifies through knowledge the world that is known.

A more basic explanation of this stance is that the reflections in these pages are, at every point, dominated by the "hermeneutic spiral." The term is taken from Ray Hart's *Unfinished Man and the Imagination: Toward an Ontology and Rhetoric of Revelation,* and refers, for the author as well as Professor Hart, to "the relation between knowing and not knowing." That is, to the question about how we *come to know something.* If we *absolutely* do not know, how are we ever to know that we do not know? If we "relatively" do not know (i.e. if there is *something* which we know we do not know), how will we ever go about adapting our research in order to find it out? The answer can only lie in some kind of "pre-acquaintance" with what there is to be known (i.e. ultimately, with the totality of being and with our relationship to it). This may still seem a rather cryptic statement. In view of the fact that it has been the methodological principle at work here, a few words of explanation are in order. Hart avoids the term "hermeneutic circle," so widely in use, first because it commonly refers to the problem of interpreting written texts, and second because it tends to connote a "vicious circle." The "hermeneutic spiral," on the other hand (imaged as a rising *helical* spiral) refers to the relation between being and knowing in which, to quote Hart's clear statement):

> The relation is circular because knowing presupposes being (minimally that of the knower and that of the known) and being presupposes knowing (at some level of apprehension of some dimension of being's givenness). Yet this is not a true circle, and therefore not a cognitively vicious circle. The circuit 'beginning' (Who can say where the circuit begins?) with the point marked 'something being given' does not return to the same point on the same circumference, but to a point 'above' it; and likewise with the point marked 'knowing.' Every spiral involves an enriched cognition of the field of the something-given, and every expansion of the field furnishes impetus for another cognitive circuit: thus an expansion of the knowing mind and of the given itself. Without some such model as that of the helical spiral to adumbrate their nature, being and knowing stand over against each other as blocks of stuff never to have intercourse, as though the givenness of being were never enhanced through knowledge (or constricted through ignorance) and as though what knowledge responds to is an incarnation of itself.[12]

Hart's application of this metaphysical statement to the practical domain of theological hermeneutics, for example, permits him to criticize the Bultmannian effort at "demythologizing" the biblical message as a mere translation of that ancient revelation into the categories of Heideggerian anthropology—without an *increment of understanding,* and hence as a non-theological enterprise. In the context of the present task of analyzing models of man, an application of the use of the "helical spiral" image has been understood in the following way: understanding involves a description of the known (concretely, an understanding of who or what man is involves a "scientific investigation" of the man-to-be-known); but description is only possible on the basis of prior knowing (concretely, we either know already who or what men is, or "science" tells us who or what we can and must look for if we are to be truly scientific). We are saved from the "which comes first" kind of question by the acceptance of the model of the hermeneutic spiral.[13]

[12] Hart, *op. cit.,* 61-62.

[13] I have given an example of this in an article, "The Phenomenological Implications of Non-Directivity," *Science et Esprit,* 20, Jan.-April, 1968, 96-111.

Second, our basic postulate is that metaphysical considerations (at least options concerning the nature and reality of the self, the other, and the world) are already at work in all human enterprises. "Metaphysics" is not a special science to be practiced by metaphysicians; if they exist as such it is only as more lucid representatives of the entire human species, natural metaphysicians.

Now that the discussion has returned to the level of metaphysical presuppositions (options), we might reflect upon the "we are" in relation to "the world." This is the third panel in the tryptich of necessary presuppositions in the philosophy of man, whether that philosophy proceeds from a systematic metaphysics of the whole of reality, even of "Being and God," or whether its process is more modestly descriptive and empirical. Here the reader should be warned that the option taken in this book for a "phenomenological" approach in preference to a systematic and deductive "metaphysical" one does not regard as even approximately adequate the current impoverishing reduction of phenomenological method to a kind of descriptive naturalism, much less a kind of descriptive psychology. It is concerned, ultimately, with the "*a priori* conditions of possibility" of the human sciences in question, but takes for granted that the "transcendental reduction," once operated, need not be continually repeated. A transcendental focus upon reality and science forever inhibits the reductionist exclusiveness of *each* science about man to designate him as "nothing but" something else.

If, then, Being appears, and if the truth of Being is its manifestation (in Heidegger's sense), an attitude toward "the world" is implied which contradicts the thesis of representational thought, in which the medium of knowledge is *Vorstellung*, representation, the mirroring of a reality already-out-there-to-be-known. The whole effort of the phenomenological movement is based upon an assertion of the grave error we commit in thinking that the world is primordially an object; thanks especially to Husserl, we may assert that the subject-object dichotomy is no more relevant on the cosmological level than it is on the psychological or epistemological level, and perhaps it is even less so. The point of view maintained here is simply that the philosophical age of representative thought has achieved all it is capable of, that the outrage of "objectivity" must be at least relativized, that at least we must remember that it is subjects who have created the world in its "objectivity." This new perspective considers the world, then, as something to be *done* as much as it is to be *known*, a *task* as much as a *given*. Human behavior, including the act of knowing or sciencing, can be defined as an organized and meaningful response to a situation which is already, but only to a degree, meaningful.

Here, some care will be necessary in order to be clear; for epistemologically the reader may be tempted to fall back into a sterile nineteenth-century idealist position and "construct" the world, and psychologically and ethically to espouse a current and popular "existentialism" in which the creation of the universe can be an arbitrary matter.

There is something to be said for both these points of view, however, provided that something is said with discretion. The present proposal is to define human consciousness, with Husserl, not as a thing but as a giver of meaning, and to define man in a preliminary way not as a "rational animal" but as essentially project and as incarnate freedom. The intention in doing so is to surpass both idealism and existentialism by subjecting both consciousness and project to a reality principle, and to make an incorporation of the non-voluntary and the non-sense (or contradiction and alienation) in the human situation an essential part of the incarnation of human liberty. The questions, "Are we *completely* free?" and "Are we *utterly* determined?" are both nonsense. The meaning given to the world is thus man's meaning, and the creation of the world by man is the creation of a human universe, in the sense of both a community of human persons and a human landscape. The "we are," then, who are in the world to begin with, find our meaning already in the world; and we define our project as part of an emergent humanity. With this perspective, I think, the *de facto* intention of both project and emergence can be determined (I do not say it is an easy task!), and deviations from the hopes incarnated in the project, whether due to fault or fallibility, can be uncovered.

There are theological and ethical corollaries to this thesis, which may be set aside for future elaboration; the main corollary in this context is epistemological, indicating that knowledge, as well as behavior, is primarily social in character. The empiricist *tabula rasa* theory is both unfruitful and misleading, belonging to a context in which absolute beginnings were thought possible.

SELECTED ADDITIONAL READINGS

FOR CHAPTER VII

Bochenski, Innocentius M., *The Methods of Contemporary Thought,* trsl. Peter Caws, New York, 1968.

Bonner, Hubert, *On Being Mindful of Man,* Boston, 1965.

Byrne, Edmund F., and Edward A. Maziarz, *Human Being and Being Human: Man's Philosophies of Man,* New York, 1969.

Coreth, Emerich, *Metaphysics,* trsl. Joseph Donceel, New York, 1968.

Crosson, Frederick J., "Phenomenology and Realism," *International Philosophical Quarterly,* 1966 (6), 455-464.

Donceel, Joseph, *Philosophical Anthropology,* New York, 1967.

Gurwitsch, Aron, *The Field of Consciousness,* 2nd. ed., Pittsburg, 1964.

Kockelmans, Joseph J., *Edmund Husserl's Phenomenological Psychology: An Historico-Critical Study,* Pittsburgh, 1967.

Kullman, Michael, and Charles Taylor, "The Pre-Objective World," *The Review of Metaphysics,* 1958 (12), 108-132.

Lonergan, Bernard J.F., *Insight: A Study of Human Understanding,* New York, 1956.

—, *Collection,* ed. F.E. Crowe, New York, 1967.

Mann, Jesse A., and Gerald F. Kreyche, eds., *Reflections on Man: Readings in Philosophical Psychology from Classical Philosophy to Existentialism,* New York, 1966.

McLeod, R.B., "The Phenomenological Approach to Social Psychology," *The Psychological Review,* 1947 (54), 193-210.

Muck, Otto, *The Transcendental Method,* trsl. W.D. Seidensticker, New York, 1968.

Natanson, Maurice, "Causation as a Structure of the *Lebenswelt,*" *The Journal of Existential Psychology and Psychiatry,* 1960 (1), 346-366.

Rahner, Karl, *Spirit in The World,* trsl. William Dych, New York, 1968.

—, *Hearers of The Word,* trsl. Michael Richards, New York, 1969.

Schmitt, Richard, "Phenomenology and Methaphysics," *The Journal of Philosophy,* 1962 (59), 421-428.

Schrag, Calvin O., *Existence and Freedom: Towards an Ontology of Human Finitude,* Evanston, Ill., 1961.

—, *Experience and Being: Prolegomena to a Future Ontology,* Evanston, Ill., 1969.

Spiegelberg, Herbert, "Toward a Phenomenology of Experience," *American Philosophical Quarterly,* 1964 (1), 325-332.

Zaner, Richard M., *The Problem of Embodiment: Some Contributions to a Phenomenology of The Body,* The Hague, 1964.